Sustainable Living

The role of whole life costs and values

Sustainable Living

The role of whole life costs and values

Nalanie Mithraratne

Brenda Vale

and

Robert Vale

 Routledge
Taylor & Francis Group

LONDON AND NEW YORK

First published by Butterworth-Heinemann

This edition published 2011 by Routledge
2 Park Square, Milton Park, Abingdon, Oxon OX14 4RN
711 Third Avenue, New York, NY 10017, USA

Routledge is an imprint of the Taylor & Francis Group, an informa business

First edition 2007

Notice
No responsibility is assumed by the publisher for any injury and/or damage to
persons or property as a matter of products liability, negligence or otherwise, or from
any use or operation of any methods, products, instructions or ideas contained in the
material herein. Because of rapid advances in the medical sciences, in particular,
independent verification of diagnoses and drug dosages should be made

British Library Cataloguing in Publication Data
A catalogue record for this book is available from the British Library

Library of Congress Cataloguing in Publication Data
A catalogue record for this book is available from the Library of Congress

ISBN: 978-0-75-068063-9

Contents

Preface

Lifetime studies are becoming commonplace in product evaluations. However, this does not apply to buildings and related activities owing to the relatively longer period involved, which leads to uncertainty. Many books which deal with the design and construction of sustainable houses have been published. However, sustainability also depends on the post-occupation activities of the users, which have not been highlighted. This book attempts to fill that gap in knowledge through the use of results from a recently completed research project.

The book demonstrates the implications of choices the designers, developers and building-users make to achieve sustainability in the residential building sector, through an analysis that covers the full lifespan. It identifies the problems associated with current practices through a lifespan model that considers costs, embodied and operating energy use, environmental impact, and global warming potential. The model was developed based on current practices employed in New Zealand and highlights the need for a new holistic approach to be taken. By considering the full lifespan, and many items such as finishes, furniture and appliances, which are usually disregarded in evaluations, the text demonstrates the importance of material and systems selection and user behaviour. It discusses the major issues ranked based on their importance for achieving greater sustainability in residential projects and highlights those which are not in common knowledge.

The book also demonstrates the practical use of life cycle analysis for achieving best practice in construction and use of residential buildings. It provides a practical guide to designers and the general public in applying the lessons learnt to individual projects to achieve sustainability in residential buildings.

The book consists of a general discussion of issues ranked based on their importance for achieving sustainability, with case studies intended for the general reader and detailed justification of the importance of issues for the more specialised reader.

Illustrations

Acknowledgements

This research was funded by a grant from the Foundation for Research Science and Technology of New Zealand.

Hayward postdoctoral fellowship from the Landcare Research in New Zealand funded the manuscript preparation.

Part A:

Construction Industry and Resource Issues

1 Introduction

Housing trends worldwide are changing rapidly, and population is increasing. There is a pressing need to understand the current state of the consumption of energy attributable to the residential sector of the global economy, and the resultant environmental impacts which, in turn, govern the sustainability of human practices. Many books which deal with the design and construction of sustainable houses have been published. However, the sustainability of houses, owing to their relatively long useful lifetime, also depends on the post-occupation activities of the users, and this is often forgotten. Studies have found that energy use and environmental impacts during this phase could far outweigh those during the construction phase. Therefore, if countries are to fulfil their international commitment to reduce greenhouse gas emissions and the resultant environmental impact, this represents a serious gap in knowledge.

This book attempts to fill that gap in knowledge through the use of results from a recently completed research project funded by the Foundation for Research Science and Technology (FRST) in New Zealand. The book is based on the results of a study of single houses in the relatively mild climate of Auckland and demonstrates the implications of choices the designers, developers and building users can make to achieve greater sustainability in the residential building sector through an analysis that covers the full lifespan of the house. It identifies the problems associated with current practices through a lifespan model which considers costs, embodied and operating energy use, environmental impact, and global warming potential. The model was developed based on current construction practices employed in New Zealand and highlights the need for a new holistic approach to be taken. By considering full lifespan, and many items such as finishes, furniture and appliances, which are usually disregarded in evaluations, the text demonstrates the importance of material and systems selection and user behaviour. It discusses the major issues ranked based on their importance for achieving greater sustainability in residential projects and highlights those which are not common knowledge.

The book also demonstrates the practical use of life cycle analysis for achieving best practice in the construction and use of residential buildings. Although commercial buildings are different from housing in terms of scale, construction, usage, and user behaviour, etc., the discussion in the book will be relevant to the small commercial buildings which make up the greater part of commercial development, because they use similar construction technology to houses, but it will not relate well to city centre high rise buildings. It provides a practical guide to designers and the general public in applying the lessons learned to individual projects to achieve sustainability in residential buildings.

The situation

From ancient times people have reacted to the natural environment and, using an acquired ability to manipulate building materials, have created a built environment not only to offer protection from the vagaries of the weather but to express an understanding of the world. Though these traditional constructions, which used materials and construction methods locally available in the vicinity, were in harmony with the environment, and were part of a sustainable environment for very many years, the conditions under which these were effective have now changed to a point where these traditional methods no longer seem wholly appropriate. The relationships between the way a particular society lived, what was built, and how it was built were interdependent relationships.

As an example, the small cottage with minimal window openings in a temperate climate was an ideal building for a society that was largely agricultural and where workers spent long hours in the open air. Home was a retreat and a place for sleeping once it was dark, so there was little need for natural light within the building. Now the reverse is true and in many cultures people spend all day away from the house in another building, or place of work. However, they expect the benefit of light when they are home in the evening, and so energy has to be expended to provide this light. The use of natural ventilation is another simple practice that has suffered because of the need for people, and often both adults in a family, to work, to support a household, with children also spending their waking hours away from home in school and after school activities. The old-fashioned wisdom was that rooms should be ventilated and that the best time to do this was midday when temperatures are highest and relative humidity is lowest. Such ventilation removed moisture from the dwelling as well as refreshing the air. Now, with the family away from home all day, windows cannot be left open for ventilation because of security concerns and in the evening it is too cold to ventilate the house by opening them. Therefore artificial methods have to be found to ventilate the building, again with expenditure of energy, because of the change in lifestyle. Rather than allowing climate and situation to create a way of living, as happened in the past, energy (mostly fossil fuel energy), is used to overcome problems created by an imposed way of life. Current society has not only moved away from environmental determinants of behaviour but also expects to be able to support a chosen way of life within a market economy independent of climate or access to resources.

Different ways of living and attitudes to home ownership can also have an effect on the choice of materials for house construction. Where dwellings are seen as a long-term investment the materials that go into making up that investment and the maintenance methods used have to be appropriate. In Singapore, the government initiative to provide affordable housing has produced a situation wherein lower middle class people can buy their own apartments. In order to make the dwellings affordable, normally in apartment blocks, the Singapore government has become a landowner, buying land for a price related to its zoned or actual use rather than its potential use in the free market. Moreover, there are policies in place for upgrading the dwellings, again subsidised, so that residents do not have to pay the full cost (Yuen 2005). The aim of government intervention is to provide all citizens with access to decent housing. As the government is the major provider of housing, the Housing and Development Board in Singapore[1] has been able to set up systems in place to either produce material, in the case of bricks, tiles, sand and granite, or to manage production of the materials required.

This approach to durable and affordable material acquisition could be contrasted with the experience in China, where the escalation in land values means that estates of apartment blocks are demolished prior to the end of their useful life. In the 1990s, along with other reforms, housing provision became part of the free market. Since then local authorities have been encouraged by developers to demolish old buildings in urban areas. While developers are able to provide increased floor area on these vacated sites, the local authorities view new modern buildings in the city centre as a boost to the local economy. As most buildings are demolished prior to achieving their full life expectancy, which is 50 years in China, it is the environment that loses out in this instance. Life expectancy of apartments in urban areas is now around 30 years and the resources used for construction of these apartment buildings have now become a waste disposal problem (Chen 2005: p. 54). It is hard to conceive of any developer wishing to invest more on durable materials in an attempt to make apartment buildings more sustainable with such a short projected life.

The same phenomenon can be found in societies where many people have their own house, while housing turnover is also vigorous. This is the case in New Zealand where the average time spent living in a particular house is around 7 years. With the thought of moving in the near future there is little incentive for home owners to invest in sustainable technologies such as solar water heaters or additional insulation as they cannot foresee any significant return on such investments. A large backlog of maintenance, which could be risky in terms of resources already contained in privately owned housing in New Zealand, has also been linked to this situation in the past. However, recent research suggests that houses lived in for less than 7 years were in need of less maintenance than those which had been lived in for longer.

> *It seems that the longer that we stay in the same house, the more likely that the house will be in the worst category* [in terms of maintenance required]. *This may relate to owners moving before conditions deteriorate, a fall-off in renovation effort with continued occupancy, or to renovation performed by the previous owner*
>
> Clark *et al.* 2005: p. 19

Here the use of housing as an investment may be encouraging better use of resources because of the deisre to maximise on the investment in the house before moving on. These examples show that the life and the resources used for housing may be controlled to a far greater extent by government policies and the norms of society than by the choices made by designers at the time the dwellings were constructed.

To keep up with the progress in social, economical, environmental and technical knowledge the construction industry has been subjected to many changes. Many studies have shown that the quality of the built environment has a direct impact on human physiological and psychological well-being (Olgyay 1963; Boardman 1991). At the same time, there is continuing pressure to create an environment conducive to higher levels of comfort, and this, together with the increasing population, means that society expends higher levels of energy than it did in the past. As buildings use more energy to maintain higher levels of comfort, efficiency improvements have tended to move along with the increased comfort expectations, but although buildings constructed today are sometimes more efficient in terms of resource use than many built two to three decades ago, still they place undue demands on the Earth.

Environmental effects of energy use

Energy issues have been a growing concern since the 1970s. Initially the main concerns were the depletion of resources and the security of supply. As a result, developments in nuclear power generation, and an emphasis on energy conservation (which is to use less energy and to accept a lesser degree of service) including measures to improve energy efficiency (which is to use less energy and enjoy the same service) of buildings such as the greater use of thermal insulation, emerged.

Over subsequent years however, the main concern has changed to the problems related to global warming and climatic change. During the early 1980s many research papers (Bach 1980; Hansen et al. 1981; Lovins 1981) were published on these subjects. Greenhouse gas emissions are now considered to be contributing to global warming (Houghton et al. 2001). Carbon dioxide, which is generated by burning fossil fuels, clearing land, making cement from limestone, etc., all of which are activities attributable to buildings and development, is one of the principal greenhouse gases which contribute to global warming. The Intergovernmental Panel on Climate Change (IPCC), which was formed in 1988, has estimated an increase in mean global temperature of $0.6 \pm 0.2°C$ during the twentieth century and projected that this increase would be $1.4-5.8°C$ by the year 2100 (Houghton et al. 2001; Trenberth 2001). Although this may not seem much, it is well above anything that the ecosystems of the earth have experienced for many centuries. Such an increase could change the wind patterns, frequency and intensity of storms, rainfall and other aspects of climate. In New Zealand, there is evidence for both progressive warming of $0.6°C$ and rise in sea level of 10–20 mm over the last century although it has not been proven to be an outcome of global warming (MfE[2] 2001: pp. 6–7). Owing to the complex nature of the earth's climatic system and the uncertain picture of the direct impact of human activity on the biosphere, the term 'global warming' has tended to be replaced by the term 'climate change'.

Climate change

Bell et al. (1996: p. 19) argue that the effects of increased carbon dioxide concentrations on the climate comprise direct and indirect effects which modify the outcome of each other. The direct mechanisms have been identified as:

- effect of biomass die-back owing to increased temperatures, which releases additional carbon dioxide;
- effect of reduced snow and ice cover in the poles owing to initial warming, which reduces the fraction of solar radiation reflected at high altitudes; and
- effect of release of methane from methane hydrates in deep oceans owing to warming which follows initial anthropogenic release of carbon dioxide.

Indirect mechanisms have been identified as:

- effect of rapid plant growth owing to high concentrations of carbon dioxide in the atmosphere, which reduces initial concentration of carbon dioxide; and
- effect of increased evaporation of water, which could assist in the formation of clouds of the correct type which in turn could increase the reflectivity of the atmosphere, thereby reducing the initial warming.

Since these positive and negative effects and their interaction have not yet been properly understood, the final outcome in terms of the system is hard to estimate with reasonable accuracy. The level of uncertainty increases when the possible direct and indirect impact on human society is considered, owing to the ability of human beings to adapt to and modify their surroundings. Some adaptations could exacerbate the problem: if temperatures rise, an increased use of air conditioners to maintain comfort would lead to increased burning of fossil fuels and escalation of the problem.

Because of these problems, current global environmental policy, such as it is, is based on the understanding that:

- human activities are currently emitting carbon dioxide into the atmosphere at a rate higher than can be absorbed by the oceans and sinks, thereby increasing the atmospheric concentration; and
- the effect of this increased concentration on the global climate is uncertain but could be harmful to all or part of the human population (ibid.). However, the next IPCC report due to be released in 2007 is expected to confirm that the risks may be more serious than previously anticipated (Adam 2006).

The New Zealand Climate Change Programme established in June 1988, explored the possible impacts of climate change on New Zealand. To bring these global issues down to the scale at which they impact directly on the lives of ordinary people, the significance of these changes to housing in New Zealand (and elsewhere in the world) can be summarised as follows.

- **Temperature changes**: An increase in average and extreme summer temperature.

  Although a global increase of 1.4–5.8°C in average temperature is expected by the year 2100, for New Zealand the increase is expected to be only about two thirds of this as the climate of the small land mass of New Zealand is influenced by the large water bodies of the South Pacific and Southern Ocean. While this increased temperature in temperate regions could lead to reduced space and water heating demand during winter, the incidence of summer overheating could increase for houses that have not been designed with adequate passive solar shading features. Any warming in tropical and subtropical areas could lead to higher demand for space conditioning throughout the year. Therefore, incorporating features such as additional insulation and shading for windows to prevent unwanted sun penetration, provision of natural ventilation systems, etc. could be beneficial for maintaining comfortable internal temperatures in summer. The energy requirement for water heating – which is a significant component of the operating energy of NZ (New Zealand) houses – is expected to decrease by 3% per 1°C increase in temperature, owing to the higher temperature of the incoming cold water (Camilleri 2000: p. 17). However, higher temperatures could also reduce the lifetime of building materials such as plastics and surface coatings used in houses, leading to shorter replacement cycles and hence increasing the energy contained in building materials in the house through the need for extra replacement components.

- **Rainfall changes**: An increase in both amount and intensity of rainfall over temperate regions with a decrease in amount of rainfall and increase in drought over subtropical regions.

Severe changes to regional climate patterns such as Asian monsoons and El Nino southern oscillations could increase the incidence of severe flooding and drought. These extreme climate events could lead to more frequent severe flooding, landslides and soil erosion, which could increase the damage to buildings, leading to an increase in flood insurance premiums or even to withdrawal of insurance cover. Increased damage to the built environment again requires more resources for replacement settlements.

- **Sea level rise**: An increase in the mean sea level.

Owing to the thermal expansion of sea water and melting of glacier and polar ice, the global mean sea level is expected to increase by up to 84 cm by the year 2100, although for New Zealand this increase is expected to be about 34 cm (MfE 2001: p. 14). A rise in sea level could lead to the obvious problem of increased coastal flooding and foreshore erosion, but, in addition, the rising water table may also reduce the capacity of existing drainage systems, causing inland flooding and hence damage to foundations and walls of houses, with a consequent further drain on resources.

While the most visible environmental damage in European and North American regions could be said to have occurred prior to 1980, for Central Asian, Far Eastern and Middle Eastern regions this has taken place since 1970 as indicated by the trend in total CO_2 emissions (see Figure 1.1). Consumption of fossil fuels such as oil, gas and coal either directly by individuals (as fuel for cooking, heating, lighting or transport) or indirectly (as fuel for manufactured products and services) can cause environmental damage, either on a local scale, for example in the form of increased air pollution in urban areas, or on a global scale through the contribution to global warming. In 1996, the resource consumption of the average person living in the industrialised countries was 4 times that of the average person in lower income countries (Loh 2000: p. 1). The disparity in the carbon dioxide emissions of various regions of the world in 1996 is shown in Table 1.1.

Recently there has been a surge in development activities in China and India, the two most populated countries of the world with a combined total of around a third

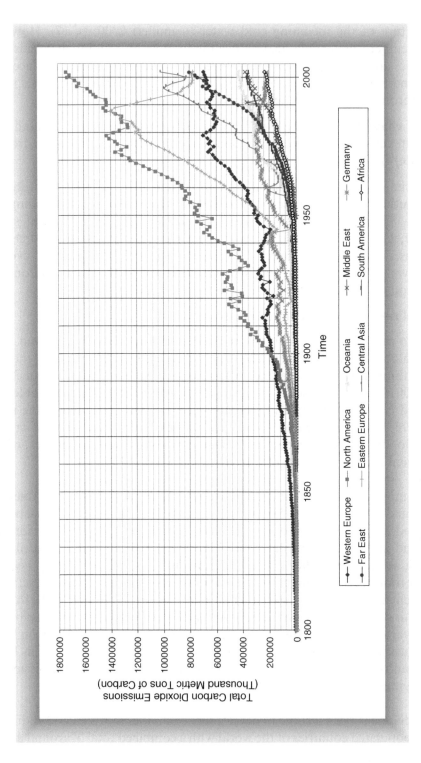

Fig. 1.1 Total global CO$_2$ emissions by region from 1800 to 2002.
(Based on: Marland, G., Boden, T. A. and Andres, R. J. (2005) Global, Regional, and National Fossil Fuel CO$_2$ Emissions. *Trends: A Compendium of Data on Global Change*. Carbon Dioxide Information Analysis Center, Oak Ridge National Laboratory, US Department of Energy, Oak Ridge, TN, USA. Available at: http://cdiac.esd.ornl.gov/trends/emis/em_cont.htm [accessed 8 March 2006].)

Table 1.1 Carbon dioxide emissions by region in 1996

Region	Population (millions)	CO$_2$ emissions (tonnes/person/year)
Asia/Pacific	3,222	2.3
Africa	710	0.9
Latin America & the Caribbean	484	2.3
Western Europe	384	9.0
Central & Eastern Europe	343	7.7
Middle East & Central Asia	307	4.0
North America	299	19.0

Based on: Loh, J. (ed.) (2000). *Living Planet Report 2000.* World Wide Fund for Nature (formerly World Wildlife Fund), p. 14

of the global population. Although, as a result, their consumption per person has not yet increased to the levels enjoyed by the rich industrialised countries, owing to the higher numbers of people involved this development could lead to greater environmental damage. Total CO$_2$ emissions resulting from energy use and per capita emissions in 2003 for selected countries are as shown in Table 1.2.

Climate change is a global problem caused by regional activities and therefore has to be tackled through local action. At the 1992 United Nations Framework Convention on Climate Change (UNFCCC) the precautionary principle was adopted and it was decided that the countries represented should aim at reducing carbon dioxide emission rates. Since this was insufficient to bring about the necessary changes, in December 1997, the Kyoto Protocol was adopted. According to the Kyoto Protocol, developed countries which have signed up to it are legally bound to reduce their emissions to set targets. Many countries, including New Zealand, have agreed to the Kyoto Protocol, which came into force in February 2005, although some important countries have not signed up to it, such as the USA and Australia.

New Zealand's greenhouse gas emissions

Although the total greenhouse gas emissions of New Zealand are comparatively small, (around 0.5% of global emissions) by the year 2003 emissions were 22.5% above their level in 1990 (MfE 2005: p. 11). According to the Kyoto Protocol, New Zealand must stabilise its emissions at the 1990 level, on average, during the period 2008–2012.

Table 1.2   Total CO_2 emissions due to energy use and per capita emissions in 2003

	Total (million metric tonnes)	Intensity (tonnes/person/year)
OECD countries		
New Zealand	38.46	9.91
Australia	376.83	19.10
France	409.18	6.80
UK	564.56	9.53
Canada	600.18	19.05
Germany	842.03	10.21
Japan	1,205.54	9.44
USA	5,802.08	19.95
Non-OECD countries		
India	1,016.50	0.96
Russia	1,503.10	11.21
China	3,307.40	2.72

Based on: Energy Information Administration (2005). *International Energy Annual 2003*. Available at: http://www.eia.doe.gov/emeu/international (accessed 3 February 2006)

The composition of the greenhouse gas emissions of New Zealand (Gg CO_2 equivalent) is shown below (See also Figure 1.2).

Carbon dioxide	46.0%
Methane	35.4%
Nitrous oxides	17.9%
Other gases	0.6% (such as hydrofluorocarbons [HFCs], perfluorocarbons [PFCs], and sulphur hexafluoride [SF$_6$])[3]

Unlike other developed countries the agricultural sector contributes close to half the emissions for New Zealand. On a CO_2 equivalent basis this sector was responsible for 49% of the total emissions in 2003. Both methane and nitrous oxides are released mainly by the agricultural sector. Emissions due to the energy sector vary widely from year to year owing to the use of thermal power stations (largely natural gas with some coal) to supplement electricity production from hydro power stations during dry years. This again is different from many other countries where there is a steady increase in the use of gas and coal for electricity generation, linked to rising living standards and possible rising populations, as discussed.

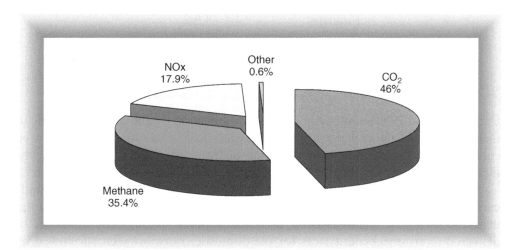

Fig. 1.2 Composition of greenhouse gas emissions of New Zealand.

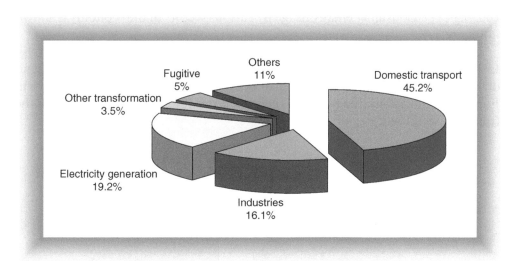

Fig. 1.3 CO_2 emissions for New Zealand by various sectors.

Carbon dioxide emissions for New Zealand by various sectors are as follows (see also Figure 1.3).

Domestic transport	45.2%
Industries	16.1%
Electricity generation	19.2%
Other transformation	3.5%

| Fugitive | 5.0% |
| Other sectors | 11.0% |

(such as commercial/institutional, residential and agriculture/forestry sectors).

Over the period 1990 to 2004, total emissions increased by about 33.8%, owing to increased use of natural gas and coal in electricity generation, and increased consumption of diesel and petrol for domestic transport (MED[4] 2005a: p. 2). The increase in emissions in the domestic transport sector since 1990 has been 61.6%. The energy sector[5] has been identified as contributing over 90% of the man-made carbon dioxide emissions in New Zealand (MED 2005b: p. 13). In 2004, the total delivered energy use increased by 2% (MED 2005b: p. 10). The main use of delivered (consumer) energy was for the domestic transport sector followed by industrial activities. During the period 2003 to 2004, the commercial, residential and transport sectors increased their share of consumer energy by 0.6%, 0.4% and 0.1% respectively, while the agricultural sector decreased by 0.9% (MED 2005b: p. 11). Consumer energy usage by various sectors of the economy is shown in Figure 1.4.

In addition to the amount of energy consumed, carbon dioxide emissions depend on the type of fuel used. Composition of delivered energy by fuel type is shown in Figure 1.5. Emissions due to liquid fuels are a result of the steady growth in the transport sector since 1990.

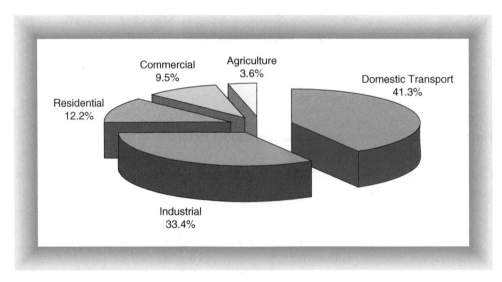

Fig. 1.4 Consumption of consumer energy by various sectors of the New Zealand economy (2004).

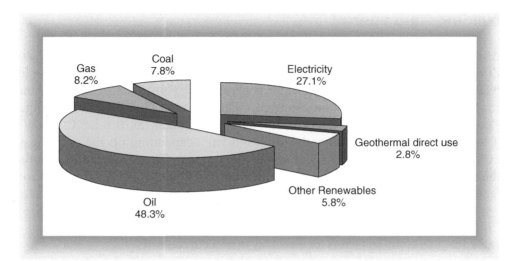

Fig. 1.5 Consumer energy use by fuel type in New Zealand (2004).

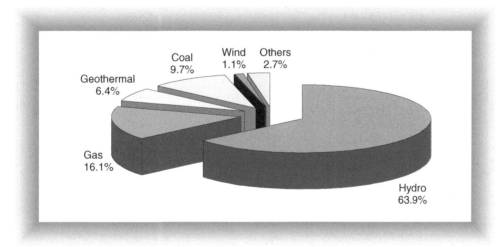

Fig. 1.6 Electricity generation by fuel type in New Zealand (2004).

While emissions due to electricity generation depend to a large extent on the proportion of hydro generation, which is influenced by the rainfall, thermal electricity generation (using oil, coal and gas) has increased to 26% of total electricity generation, increasing the carbon dioxide emissions in the year 2005 (MED 2005b: p. 120). The use of coal for electricity generation has also increased from 25 to 38% of thermal generation owing to the recent depletion of the Maui gas field. Wind generation is starting to play an increasing role in the electricity generation mix with a contribution of 1.1% to the renewable sources (ibid.) (see Figure 1.6).

Buildings and energy

About 12.2% of the total energy and 34.2% of the electricity consumed in New Zealand are used in the residential sector (MED 2005b: pp. 11 and 121). With its relatively high percentage of renewable generation (mostly hydro), New Zealand can be said to have 'low-carbon' electricity. The CO_2 emissions factor of 'average' electricity generation in New Zealand is usually taken as 0.1 kg CO_2/kWh (Camilleri 2000: p. 58), although this could vary depending on the amount of available hydroelectric capacity at any given moment. The equivalent factor for thermal electricity is 0.64 kg CO_2/kWh (ibid.). Table 1.3 shows the emission factors for a range of other countries as a comparison. (Fuel mix used for

Table 1.3 CO_2 emission factors for electricity in selected countries

Country	Fuel mix	CO_2 emissions factor (kg/kWh)	Information sources
UK (2005)	Natural gas (39.3%) Coal (33.4%) Nuclear (20.6%) Renewables (3.8%) Other (2.9%)	0.46	Electricityinfo.org. (2006).
Australia (2005)	Black coal (54.8%) Brown coal (21.9%) Gas (14.2%) Hydro (6.8%) Oil (1.3%) Other (1.0%)	1.051	Australian Institute of Energy (2006) and Uranium Information Centre (2006).
France	Nuclear (78%) Hydro (11%) Fossil fuels (10%) Solar/wind (0.2%) Other (0.6%) (2003)	0.056 (1995)	International Energy Agency (2004) and de Bustamante, (2006).
Japan	Fossil fuels (64%) Nuclear (23%) Hydro (11%) Geothermal (0.3%) Other (2.4%) (2003)	0.44 (1998)	International Energy Agency (2004) and US Environmental Protection Agency (2006).
Taiwan (1999)	–	0.7	US Environmental Protection Agency (2006).
Thailand (1999)	Gas Coal Fuel oil	0.75	US Environmental Protection Agency (2006).

electricity generation is highly variable between years. The fuel mix and emissions values shown in the table may not be for the same year.)

Although this relatively 'low carbon' electricity is the main source of the energy used in most NZ houses, carbon dioxide emissions attributable to the residential sector of the New Zealand economy could be considerable. In addition to emissions due to operating energy use, certain building materials used in residential buildings also contribute significantly to carbon dioxide emissions. Apart from the energy sector and domestic transport, industrial production of steel, aluminium, cement, lime (attributable to the building industry) and urea is the other main source of carbon dioxide emissions in New Zealand (MED 2005a: p. xix).

According to the Building Research Establishment of the UK (Rao et al. 2000: p. 8), about 50% of the UK's greenhouse gas emissions are attributable to the energy used in buildings, while in the USA, buildings account for 48% of the total energy use (AIA 2005). Although these figures are based on the operating requirements of the buildings, Australian research (Fay 1999) has highlighted the importance of both the operating and embodied energy attributable to buildings. This suggests that construction choices could be of significant importance in terms of the total energy attributable to buildings. Therefore for a holistic approach, the total energy used by the building throughout its useful life, known as life cycle energy, has to be considered. Life cycle energy attributable to buildings consists of energy embodied in the materials and elements used in construction – known as initial embodied energy; energy to operate the building throughout its useful life – known as operating energy; and the energy added during the maintenance, renovation and replacement of the building materials and elements – known as recurrent embodied energy. There are important interactions between initial embodied energy, recurrent embodied energy and operating energy, and trade-offs are possible between the three categories; for example, while the use of a durable material may lead to a higher initial embodied energy it may also require less maintenance and therefore lead to lower life cycle energy. Similarly the use of high levels of insulation or multiple glazing will increase the initial embodied energy, but will reduce the overall operational energy demand. For a full life cycle assessment the energy to dispose of the materials of a building at the end of its life and return the site to a natural condition should also be considered.

Urban development and residential constructions

Urban intensification in terms of medium density (low rise apartments and terraced/town houses) and high density (mid to high rise apartments) developments has been the global norm since the 1990s in an attempt to curb urban sprawl. Efforts have also been made to create pedestrian-frendly high density housing with more opportunities for social interaction and stronger communities. These more intensive developments have also been considered to be more sustainable in environmental and economic aspects although the accept-ance of intensification is being questioned by empirical studies (Jenks et al. 1996; Talen 1999).

Different parts of the world have different housing forms representing the norm, but the same issues of materials and environmental impacts can be seen. In China, for example, about 78% of the urban population live in apartments, with low density housing forming only 1% of the total. As a result of housing reforms in the 1980s, per capita living space increased from 3.6 m^2 to 10.3 m^2 by 2004. Clay bricks, which were a traditional building material for these develop-ments, were banned in 2000 in an attempt to curb air pollution problems caused by the use of coal as a fuel for firing the bricks. Also, traditional timber frame construction has been replaced by more affordable concrete and steel (Robbins et al. 2004). (Anecdotal evidence suggests that developers in other countries, espe-cially the USA, are looking for cheaper alternative structural systems as a result of the high price of steel driven by the construction sector in China and India.) Poor quality of construction and substandard building materials are a concern for apartment buyers in China. The standard practice has been to sell 'shell' homes with unfinished concrete walls, floor and ceilings and without any fittings, which would then be completed by the home-owners themselves or by using cheap labour leading to unreliable work with high wastage. During the period 1986 and 1997 the furniture industry in China had an annual growth of 40% (Luo and Perez-Garcia 2001). Since 2000 a regulation change has required developers to provide fully fitted homes. Although this could reduce some wastage of resources, apartments in China have also been increasing in size, a situation that is found in other countries. In the USA, the houses themselves have also become significantly larger, increasing from an average of 90 m^2 in the 1950s to 211 m^2 in 2004 (Glink 2005). Thus, at a time when there is concern about the environmental impact of fossil fuel use, houses around the world are increasing in size, and thereby drawing on more resources for their construction.

Housing in New Zealand

The building and construction industry is a major sector in the New Zealand economy, and it directly employs around 6% of the workforce with many more indirectly involved in support industries. Construction of houses makes up almost two thirds of the building and construction industry. The residential property market, with a total value of between 450 and 500 billion New Zealand dollars, is the largest investment sector in New Zealand (CfHR[6] 2004: p. 5).

The rate of home ownership is high in New Zealand and is among the highest in the OECD. However, there has been a moderate decline in the countrywide rate of home ownership between 1981 and 2001. In 2001 the home ownership rate was 67.8% (Statistics NZ 2002: p. 10). Housing is also a major form of investment and accounts for about 25% of the consumer price index (CPI). The trend in housing has been for separate and larger houses with 80% of the total housing stock in 2001 being separate houses, with three-bedroom houses being the most common type (Statistics NZ 2002: p. 9). An increase in the number of new houses in multi-unit blocks and terraced housing and apartments, in an effort to provide 'medium density housing', is the current trend in inner city locations, particularly in Auckland and Wellington. In 2003, of all residential building construction consents issued, 23% were for apartments (CfHR 2004: p. 4).

Although the number of houses increased by 6.5% between 1996 and 2001 (Statistics NZ 2002: p. 9), the population growth was slower with an increase of only 3.3% (Statistics NZ 2001). However, during the same period, there was a growth in the population of the Auckland region of 8.4%. Although this was not the highest growth rate, Auckland had the highest numerical increase. In the 2001 census, the resident population in the Auckland region reached 1.16 million, which is about one third of the population of the whole country (Statistics NZ 2001). Auckland is also the most populated region in the country. Of the New Zealand population, 10% live in Auckland City with a further 20% living in the rest of the Auckland region. In contrast to the rest of the country, Auckland has a higher than normal concentration of households with two or more vehicles.

Notes

[1]http://tcdc.undp.org/experiences/vol4/Public%20housing.pdf: p. 17
[2]Ministry for the Environment, New Zealand
[3]Global warming potentials of these gases are discussed in Chapter 5.
[4]Ministry of Economic Development, New Zealand
[5]This includes domestic transport, thermal electricity generation, and other energy transformation industries, but not international transport. Sources of energy and the pattern of consumption are discussed in detail in Chapter 3 (see Energy sources and Energy use)
[6]Centre for Housing Research, New Zealand

References

Adam, D. (2006) Climate scientists issue dire warning. *The Guardian*, 28 February 2006.

American Institute of Architects (2005) *Architects call for fifty percent reduction by 2010 of fossil fuel used to construct and operate buildings.* Available at: http://aia.org [accessed 9 January 2006].

Australian Institute of Energy (2006) *Energy value and greenhouse emission factor of selected fuels.* Available at: http://www.aie.org.au/melb/material/resource/fuels.htm [accessed 4 May 2006].

Bach, W. (1980) The CO_2 issue: What are the realistic options? *Climate Change* 3: 5.

Bell, M., Lowe, R. and Roberts, P. (1996) *Energy Efficiency in Housing.* Ashgate Publishing Ltd.

Boardman, B. (1991) *Fuel Poverty.* Belhaven Press.

Camilleri, M. J. (2000) *Implications of Climate Change for the Construction Sector: Houses* (BRANZ Study Report No. 94). Building Research Association of New Zealand.

Centre for Housing Research (2004) *Changes in the Structure of the New Zealand Housing Market: Executive Summary.* Prepared by DTZ New Zealand Ltd.

Chen, K. (2005) *The Sustainable Apartment Building.* Master of Architecture Thesis, University of Auckland.

Clark, S. J., Jones, M. and Page, I. C. (2005) *New Zealand 2005 House Condition Survey* (BRANZ Study Report No. 142). Building Research Association of New Zealand.

de Bustamante, A. S. (2006) *The nuclear paradox.* Available at: http://www.din.upm.es/ trabajos/amalio/index.html [accessed 8 May 2006].

Energy Information Administration (2005) *International Energy Annual 2003.* Available at: http://www.eia.doe.gov/emeu/international [accessed 3 February 2006].

Electricityinfo.org (2006) *Fuel mix disclosure data.* Available at: http://www.electricityinfo.org/supplierdataall.php [accessed 3 May 2006].

Fay, M. R. (1999) *Comparative Life Cycle Energy Studies of Typical Australian Suburban Dwellings.* PhD Thesis, Deakin University, Australia.

Glink, I. R. (2005) *Housing Trends 2004.* Available at: http://www.thinkglink.com [accessed 19 December 2005].

Hansen, J. et al. (1981) Climate impact of increasing atmospheric CO_2. *Science* 21(3): 957–966.

Honey, B. G. and Buchanan, A. H. (1992) *Environmental Impacts of the New Zealand Building Industry* (Research Report 92/2). Department of Civil Engineering, University of Canterbury, New Zealand.

Houghton, J. T. et al. (2001) *IPCC Climate Change 2001: The Scientific Basis.* Cambridge University Press.

International Energy Agency (2004) *Electricity Information.* IEA Statistics.

Jaques, R. (1996) Energy Efficiency Building Standards Project: Review of Embodied Energy. In *Proceedings of the Embodied Energy: The Current State of Play Seminar*, (G. Treloar, R. Fay and S. Tucker, ed.) pp. 7–14, Deakin University, Australia.

Jenks, M., Burton, E. and Williams, K. (eds) (1996) *The Compact City: A Sustainable Urban Form?* Spon.

Loh, J. (ed.) (2000) *Living Planet Report 2000*. World Wide Fund for Nature.

Lovins, A. B. (1981) *Energy Strategy for Low Climatic Risks*. Report under contract 104 02 513 to Federal Environmental Agency, West Germany.

Luo, J. and Perez-Garcia, J. (2001) *China's Housing Market: A Policy Assessment and Outlook for wood Consumption*. CINTRAFOR working Paper No. 83. Available at: http://www.cintrafor.org [accessed 16 January 2006].

Marland, G., Boden, T. A. and Andres, R. J. (2005) Global, Regional, and National Fossil Fuel CO_2 Emissions. *Trends: A Compendium of Data on Global Change*. Carbon Dioxide Information Analysis Center, Oak Ridge National Laboratory, US Department of Energy, Oak Ridge, TN, USA. Available at: http://cdiac.esd.ornl.gov/trends/emis/em_cont.htm [accessed 8 March 2006].

Ministry for the Environment (2001) *Climate Change Impacts on New Zealand*. Ministry for the Environment.

Ministry for the Environment (2005) *New Zealand's Greenhouse Gas Inventory 1990–2003: The National Inventory Report and Common Reporting Format Tables*. Ministry for the Environment.

Ministry of Economic Development (2005a) *New Zealand Energy Greenhouse Gas Emissions 1990–2004*. Ministry of Economic Development.

Ministry of Economic Development (2005b) *New Zealand Energy Data File – July 2005* (compiled by Hien, D. T. Dang). Energy Modelling and Statistics Unit, Energy and Resources division, Ministry of Economic Development.

Olgyay, V. (1963) *Design with Climate: Bioclimatic Approach to Architectural Regionalism*. Van Nostrand Reinhold.

Rao, S., Yates, A., Brownhill, D. and Howard, N. (2000) *EcoHomes: The Environmental Rating for Homes*. Construction Research Communications Ltd. (for Building Research Establishment Ltd).

Robbins, A., Boardman, P. Perez-Garcia, J. and Braden, R. (2004) *China Sourcebook: An Introduction to the Chinese Residential Construction and Building Materials Market*, CINTRAFOR Working Paper No. 94. Available at: http://www.cintrafor.org [accessed 16 January 2006].

Statistics New Zealand (2001) *Central Auckland zone urban area community profile*. Available at: http://www2.stats.govt.nz [accessed 26 August 2005].

Statistics New Zealand (2002) *2001 Census of Population and Dwellings: Housing*, Wellington. Statistics New Zealand.

Statistics New Zealand (2006) *New Zealand in the OECD*. Available at: http://www.stats.govt.nz [accessed 13 January 2006].

Talen, E. (1999) Sense of community and neighbourhood form: An assessment of the social doctrine of new urbanism. *Urban Studies* 36(8), 1361–1379.

Trenberth, K. E. (2001) Stronger evidence of human influences on climate: The 2001 IPCC assessment. *Environment* 43(4): 8.

Uranium Information Centre (2006) *Australia's electricity*. Available at: http://www.uic.com.au/nip37.htm [accessed 3 May 2006].

US Environmental Protection Agency (2006) *Asian electricity LCI database*. Available at: http://www.epa.gov/ord/NRMRL/Pubs/600R02041/AsiaElec.pdf [accessed 3 May 2006].

Yuen, B. (2005) Squatters No More: Singapore Social Housing. *Third Urban Research Symposium: Land Development, Urban Policy and Poverty Reduction*, 4–6 April 2005. Brasilia, Brazil, p. 11.

(2006) *Provision of public housing in Singapore*. Available at: http://tcdc.undp.org/experiences/vol4/Public%20housing.pdf [accessed 16 March 2006].

2 Life cycle analysis

In order to account for resource use and the resulting environmental impacts over the construction, use and demolition phases of houses, an assessment method is required which covers all these phases and is capable of quantifying various resource uses at different times. This chapter examines the theory of life cycle analysis, which fulfils all or most of these requirements, and its application to buildings and related issues.

Life cycle analysis (LCA) is a quantitative assessment of resource uses (raw materials and energy) and waste discharges for every step of the life of products, services, activities and technologies, and thereby provides a way to evaluate and quantify the environmental impacts of a wide range of products and activities (Krozer and Vis 1998: p. 53; Chevalier and LeTeno 1996: p. 488). The international standard on LCA (ISO 14040 1997) requires the consideration of environmental impacts under the categories of resource use, human health and ecological consequences. LCA includes the entire life, including extraction and processing of raw materials, supply of energy, manufacturing, transportation and distribution, use/re-use/maintenance, recycling and final disposal of the product.

Depending on the system boundary selected, life cycle analysis is of two kinds.

1. Cradle to grave analysis – analysis of the entire life of a material or product up to the point of disposal.
2. Cradle to cradle analysis – analysis of a material or product the life of which does not end with disposal but which becomes the source of a new product through a recycling process.

While LCA has been used since the 1960s for performance evaluations, during the 1970s, the focus of 'cradle to grave' LCA was on calculating energy requirements as a direct result of the first 'oil shock', with limited attention given to environmental issues. Since the 1980s, LCA has included non-energy-related studies (Verschoor and Reijnders, 1999: p. 375), such as those linked to clean production, process development and environmental labelling.

Hobbs (1996), quoted by Jaques (1998), has identified the following as the objectives of LCA:

- to compare alternative processes;
- to improve resource efficiency;
- to assess environmental impact;
- to identify ways of reducing the impact; and
- as a source of information on resource use and emissions into the environment.

When applied at the design stage, LCA can demonstrate which environmental impacts are the most important in the life cycle and thus guide the designer to minimise these. However, when the budget is limited, tackling the single most important impact may be all that can be achieved for a particular project, even though, ideally, all significant impacts should be addressed.

Life cycle analysis methodology

LCA methodology consists of four major steps (Jönsson et al. 1996: pp. 245–246).

1. Definition of goal and scope
2. Inventory
3. Impact assessment
4. Evaluation and interpretation.

Studies which cover only the first two steps are regarded as life cycle inventories and not life cycle analyses. All these steps are discussed below.

Definition of goal and scope

The need for LCA and the intended use of the results are established during this step. In turn, these factors can drive the design of the actual LCA. A simplified model of the product or the process is established which defines the purpose and the boundaries of the study, and establishes the functional unit, evaluation period and quality assuarance procedures. Figure 2.1 shows an example of a system boundary that may be used for a life cycle assessment of building construction activities.

System boundaries are used to simplify the analysis as it is usually not possible to trace all the inputs and outputs relating to a product or a process. However,

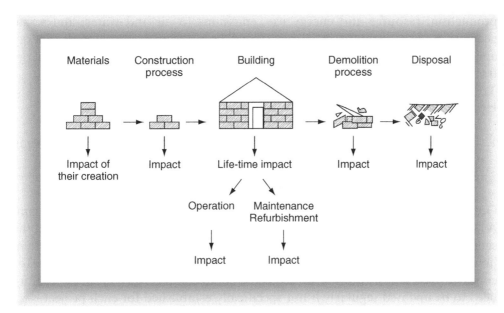

Fig. 2.1 Sample system boundary for a building LCA study.

excluding the activities beyond this established boundary can sometimes distort the results of the analysis. For example, if the energy and the impact of transporting materials to the building site are ignored, but the designer chooses to specify materials that have to be trucked long distances using an inefficient transport system, such as road haulage, the energy that goes into making the building may be underestimated. A similar example would be the issue of insurance. Specification of a difficult construction process may incur higher insurance premiums. These increased costs include a higher than normal share of the energy and resources which are required to operate the insurance company. Therefore, excluding insurance in such a case could skew the results. However, the common practice in LCA is to include all the processes during the lifetime of a product, such as a building, but in the case of capital goods to include only the environmental impacts of the production of materials necessary for the production of capital goods. A simpler approach would be to include only the processes that make a significant contribution to the overall environmental impact. However, in the case of repetitive processes, even though the contribution per process is trivial, the total contribution could become considerable.

Inventory

The inputs and outputs that occur over the life cycle of the product or process are quantified and analysed in this step. Data gathered would comprise, for example, production, resource and energy use, emissions to air and water, and waste generation. The resultant inventory would be a list of raw materials, energy and the various forms in which it is supplied, semi-finished products/energy from other processes, etc. used and the emissions that would occur.

Hence the data gathered would come under the following two categories.

1. Foreground data – those which are specific to the process or product being considered; and
2. Background data – generic data on materials, energy, transport, etc.

Often a high percentage of all data used for a particular LCA is background data. As such, only background data that have been derived based on the goal and the scope of the LCA for which the data are to be used should be selected. Often, however, generic background data are used. Use of generic data is not an issue for comparative studies, although all such data could change over time and therefore need to be updated regularly if not specifically established for each project.

Impact assessment

This step involves understanding and evaluating the significance and the magnitude of the effects of the environmental burdens identified and quantified in the inventory. This is the most complicated step in LCA owing to there being insufficient data on the exact damage to ecosystems caused and the lack of standard methods for evaluation of this damage. The choice of methods leads to wide differences in answers in situations where the same product is being assessed by different teams, as shown in Table 2.1. However, what should be noted from the table is that though the absolute values are different the order of

Table 2.1 Comparison of life cycle energy of average New Zealand house

Results	Notes	Reference
Construction energy 298 GJ	Initial construction energy use only	Baird and Chan, 1983
Annual operating energy 32.89 GJ		
Construction energy 347 GJ	Construction energy includes initial construction, demolition and disposal	Jacques, 1996
Annual operating energy 28.58 GJ		

magnitude is similar. Hence, LCA, for all its faults and uncertainties, can provide useful answers in indicating the likely environmental impat of a product.

Impact assessment is further subdivided into four steps; classification, characterisation, weighting and valuation. While classification and characterisation are compulsory, weighting and valuation are optional and are sometimes grouped together:

- **Classification** – This is used to assign impact factors identified in the inventory to various impact categories, such as resource depletion, damage to an ecosystem, global warming, acidification, etc. Several inventory categories may be assigned to a single impact category, while a single inventory category may also be assigned to several impact categories. For example, both CO_2 and methane (CH_4) could be assigned to the global warming category, while sulphur dioxide (SO_2) could be assigned to both human health and winter smog categories. However, the actual impact of emissions on the environment will not only depend on the quantity but on time and location, which determines the concentration, and these are not considered in LCA. Therefore, results of an LCA give an indication of a hazard rather than the actual risk. A hazard is a source of a risk which is independent of the location. If the actual risk is to be evaluated, both the location and the nature of exposure to the emission have to be considered.
- **Characterisation** – This is used to assess the relative contributions of impact factors and then to aggregate them within the impact categories. For example, over a period of 100 years the global warming potential of 1 kg of methane is 21 times that of 1 kg of CO_2. This could be used as the basis for deriving characterisation factors (1 for CO_2 and 21 for methane). Multiplying the impact factor result by the characterisation factor would give the impact indicator result.
- **Weighting and valuation** – This is used to derive the relevance of the inventory categories in terms of environmental implications, and is very controversial as it is often difficult to interpret resource flows without using subjective judgements. In order to evaluate the relative importance of impact categories, weighting is applied across the categories. These valuation methods can be either quantitative or qualitative. Lindeijer (1996) has classified quantitative methods into five basic categories, as shown below.

  1. proxy: use of selected indicators to represent the total environmental impact;
  2. technology: systems such as ecological footprint (used to estimate the biologically productive area required to support consumption);

3. panels: use of public opinion of seriousness of an impact to derive a default weighting;
4. distance-to-target: setting up a target for each impact category to use for weighting. The further away from the predefined target, the more serious the problem; and
5. monetisation: damage expressed in terms of some monetory unit. This approach can include the use of concepts such as 'present cost', 'willingness-to-pay' to avoid the impact (acceptance by society), and 'future extraction costs' for resources.

The most commonly used methods for valuation are distance-to-target and willingness-to-pay. However, an individual's willingness-to-pay may differ from that of society. This is illustrated by the reluctance of individuals to take up renewable technologies such as solar water heating in the home, even though such technologies could be advantageous to society perhaps in terms of helping that society to meet its Kyoto targets.

Evaluation and interpretation

This step involves interpretation of results and evaluation of the options for reducing the environmental impacts or burdens identified. LCA results generally suffer from data issues, model accuracy and model completeness issues. Uncertainty of results due to the data quality is calculated using statistical methods and expressed as a range or standard deviation. Uncertainties in the model can be evaluated by sensitivity analysis, which involves changing assumptions and recalculating the results of LCA. This would indicate the significance of a certain assumption in terms of the final result. While the result may be heavily dependent on particular assumptions used, there may be certain other assumptions which hardly make any difference to the final result. When the results/ conclusions of LCA are discussed, the defining assumptions on which the results are based also have to be included. Completeness of the model can be assessed using contribution analysis, which identifies the processes that make crucial contributions and therefore are of utmost importance to the final result. If necessary, these processes could be further analysed to identify the accuracy and completeness of representation in the model, as the final result is heavily dependent on these processes.

Therefore, depending on the system boundaries selected, LCA can either be very complex and time consuming or much briefer but much more prone to error. LCA has also been criticised for being very data intensive and for its methodological imperfections (Verschoor and Reijnders, 1999: p. 375). These imperfections arise as a result of the following assumptions inherent in the analysis.

1. Time stability – The product system is analysed as a time-stable system. This process is often referred to as back-casting technique, as it allows future scenarios to be assessed based on current knowledge. Although this allows for an instant picture of a product's life this is not realistic; for example, when the service life ends it is assumed the product is disposed of in the same way as it would have been when it first came into use, neglecting possible evolution of technology. Thus, it might be assumed that copper–chrome–arsenic treated waste timber cannot be burned for fuel as at present, although improvements in combustion technology to remove the toxins could allow this to occur in the future when a building containing such timber is demolished.

2. Separability – The process considered is assumed to be completely separated from the other products outside the system. Although this assumption avoids an unmanageable expansion of the system boundary it does not reflect the reality. Often, the manufacture of a particular product generates various by-products which may use vastly different amounts of resources. For example, during the process of squaring and dressing of timber to produce planks, large amounts of sawdust are also produced. Although sawdust can be seen as a waste product, it could also find a use in pressed boards, or even as aggregate in some concrete products. However, it is not possible to divide the process into separate phases responsible solely for planks or sawdust in order to allocate resource use and environmental impacts. If the two processes are considered separately, the results will be different from those arising from consideration of a combined process. Another instance would be a process that produces waste heat. The latter would be calculated as part of the energy consumption of the process. However, if that waste heat were to be utilised in the form of contributing to a district heating scheme, the makeup of the total energy input would indicate far greater efficiency. Separating the processes, however, might not reveal this. Normally, environmental loads created by individual products are allocated based on either mass or energy use. Economic value is another parameter used for allocation.

3. Precision – It is assumed that any flow can have only one accurate value. All values used in the process evaluation are averages for the industry, and may vary from the actual values for a product in a particular place on a particular day. Resource use for a particular situation depends on the actual working conditions, experience of those who work on the job, etc.

4. Steady state.

5. Punctual and continuous world model – All environmental flows are assumed to come from and go back to the same source.

Although LCA data are not absolute values because of the assumptions discussed above, the method provides a tool for quantification of the environmental impacts through the life cycle of the product based on comparative studies. As such it can aid in the development of more efficient and environmentally sound practices. However, for comparisons to be meaningful, the functional unit which forms the basis for comparison has to be selected carefully. Further discussion on the functional unit appears below in the relevant section.

Life cycle analysis of buildings

LCA was applied to whole buildings and building materials from the late 1970s (Jacques 1998) in an attempt to safeguard limited non-renewable resources (Bekker 1982: p. 55). While some analytical tools rely heavily on the traditional LCA methodology described above, others use a simpler method based on a quantitative/qualitative approach. Two main concerns in applying LCA to buildings and building materials are the relatively long lifetime of the product, as many changes can occur during the lifetime of a building, and the complex and often ill-defined ways in which a building may be used (Jönsson et al. 1996: p. 245). Since LCA involves the investigation of the complete lifetime prior to any decision-making, when applied to buildings (which may last from a few years to several centuries), investigation and decision making become harder as nobody can predict at which point the total replacement of a building becomes necessary in terms of technical, economical and social requirements.

As identified by Chevalier and LeTeno (1996: p. 490), in comparison with other products to which LCA is applied, building materials are unique as they are subjected to three phases: installation, service life and end of life. Installation begins when materials are brought to the site and ends when installation is complete. Stocking and moving of materials within the site that may occur between these points is more dependent on the building type and site conditions than on the building material itself. Energy consumption during this phase depends on the type of building, site planning and quantities of waste (determined by the building design and specification, and the skill of the workers). The basic LCA assumption of precision is contradictory to this situation. The service life of building materials may vary from several years up to 100 or more, depending on external conditions such as climate, type of user and change of use. Depending on the same conditions maintenance and replacement occur at varying frequencies, and are contradictory to the flow accuracy assumption of LCA. These replacement cycles can have a considerable influence on the LCA results. The end-of-life phase begins when the materials are removed from the building during demolition. After the building has been demolished it becomes difficult to determine the role played by these materials and to quantify the input and output. Here separability becomes the issue.

Further, the maintenance schedules often used at present are based on previous experience and the technological advancements in building materials and elements that may lead to more durable materials and changes in user behaviour are not considered in the analysis. Embodied energy is also assumed to remain constant over the entire lifetime of the building. However, Alcorn and Haslam (1996: p. 138) have established that the embodied energy coefficients of New Zealand building materials have decreased by an average of 41% for almost one third of these materials, while for the remaining two thirds they have increased by an average of 46%, during the period 1983 to 1996, as a result of changes and advances in material production technologies. This is contradictory to the basic assumption of time stability.

The focus of life cycle analysis has normally been the impact on the global and regional environment rather than on local or internal environments. This is contradictory to the usual intention of general building activities which aim to create both internal and external environments at the local scale. The indoor home and work environments, which have implications for occupant health, are an important aspect of buildings that could be included in life cycle evaluation. Their relevance has been debated (Antonsson and Carlsson 1995; Jönsson 2000) though they are not commonly included in building life cycle studies. The quality of the indoor environment could be argued to be the consequence of the use made of the building or even the materials/elements that are used in its construction and is a continually changing parameter. The qualitative and semi-quantitative nature of indoor environmental condition indicators such as volatile organic compounds (VOCs) and indoor air quality (IAQ) and the general use of the physical building as the functional unit are the reasons for neglecting local and indoor environments in most life cycle studies. In any case, Barnthouse et al. (1998) have argued that inclusion of local and indoor environmental impacts is problematic owing to increasing inaccuracies resulting from the LCA methodology. This is a result of the utilisation of an aggregate environmental impact over the various phases of building life without consideration of time and place (concentration). This gives rise to a poor relationship between the predicted/potential impacts and actual impacts in local or indoor environments.

Studies which include the infrastructure essential for the successful operation of buildings, such as waste water disposal systems (Erlandsson and Levin 2005; Wu et al. 2005; Zhang et al. 2006), or roads and parking facilities (Li 2005) are also very limited.

Functional unit

For meaningful comparative analyses to be made, the products and services being considered should fulfil the same function. For example, a disposable battery which is thrown away after a single use cannot be compared with a rechargeable battery which may be charged and used many times. However, a comparison could be made based on a similar number of hours of use of either product. For example, 100 hours of use may be achieved with either five disposable batteries or one rechargeable battery charged five times. Hence the energy and resources it takes to achieve these two actions could be compared.

Buildings also have several functions and, therefore, the functional unit in a building-related LCA has to be defined so that compared buildings perform the same service, such as houses, office buildings, hospitals, etc. Usually the whole building, which is comprised of building elements that are in turn also composed of several material layers with definite characteristics, is considered as the functional unit. The system boundary used would depend on the phase of the building being considered. While material production is usually considered from raw material extraction to the factory, the use phase of the building is normally limited to the property boundaries. However, depending on the scope of the LCA, the study boundary could be extended to include building occupant support systems such as water supply, waste disposal, and transport.

Comparative analyses of building types sometimes use unit floor area as the functional unit. Although this could provide useful data, such comparisons usually tend to imply that the larger building is superior in performance, owing to the economy of scale deriving from larger spaces being enclosed by the envelope. In fact, the overall impact of a large building tends to be greater as a result of the additional resources necessary for construction and operation. Other common functional units used are a bedspace or a room, in hotels, hospitals, etc., and a workspace, in offices, laboratories, etc..

Useful life of buildings

The longer the useful service life of the building, the less will be the burden on the environment and limited natural resources. However, although the age and the state of the building may affect its ability to provide the minimum acceptable services required, these are not the governing criteria of the useful life of a building. The increase in the demand for new buildings can be inversely proportional to the actual useful life of the building. The useful life is determined by a series of environmental, structural, user-specific and economic factors. On the other hand, the useful life of a building may also depend on its architectural and historical merit. While environmental factors such as climate, temperature ranges, moisture and biological factors such as fungi, bacteria, etc. affect the useful life of a building, user-specific factors such as stresses imposed by human beings while the building is in use also contribute to deterioration, and the higher the deterioration the faster tends to be the eventual destruction. Even though, theoretically, an increase in the useful life can reduce the use of energy and raw materials, in reality, with increased time, maintenance may become a burden and the consumption of energy required to maintain the necessary comfort conditions could become too high. An example would be deterioration in window joinery leading to greater operational energy use owing to infiltration through leaky joints. Therefore, the useful life depends on the fitness of the building for the intended use.

The fitness of the building is dependent on a series of negative and positive effects. When the negative effects take over, the useful life comes to an end. Bekker (1982: p. 57) identified the following as determinants of the useful life.

- economic effects – cost;
- physical effects – consumption of energy, raw materials, use of land and space;
- social effects – income, living standards, health aspects;
- ecological effects – pollution, loss of ecosystems.

Generally, buildings are demolished when they reach the end of their economic life, which may be much shorter than the physical life determined by the structural system. Economic life is defined as the period between construction and the time when the market value of the property inclusive of the cost of demolition and clearing the site is less than the value of the cleared site for new use (Johnstone 2001: p. 43).

Because of these factors, LCA is often carried out over a predetermined, assumed lifetime. Although this does not represent the true life of the building, some aspects of LCA, such as simulations of annual energy use, if done over a long period become increasingly meaningless owing to unpredictability in future energy costs and comfort expectations. Nonetheless, using a short building life undermines the potential long-term benefits of embodied energy in life cycle terms. If the building is demolished prior to completion of its useful life, energy embodied in the materials is wasted. The REGENER project (1997) estimated the lifetime of buildings in Europe to be 80–120 years, and recommended that LCA should cover up to 100 years, although 80 years was used for its reference house. Adalberth (1977) used 50 years for a Swedish house, Fay (1999) 100 years for an Australian house and Zhang et al. (2006), 50 years for a Chinese house as the useful lifetimes for their respective life cycle studies.

Useful life of an NZ House

Although no standard convention exists for the useful life of an NZ house, consistent with the requirement of the New Zealand Building Code that the serviceable life of residential buildings be 50 years, Jacques (1996) assumed a life of 50 years in his work. Although it is not clear how they arrived at their conclusion Wright and Baines (1986: p. 69) used 80 years in their study arguing that 'houses in New Zealand last something like 80 years', while Johnstone (2001: p. 48) has estimated the current useful life of an NZ house to be 90 years.

As LCA attempts to provide a broad assessment of the environmental impacts associated with a product during its whole lifetime, application of LCA to buildings can become a tedious and time-consuming process. There are also certain contentious issues to deal with, such as the impact of the loss of land resulting from building activities, which is difficult to assess as it differs with cultural background. As a result, many LCA tools use a simpler approach that combines quantitative and qualitative features to evaluate environmental issues.

Use of natural resources such as water and energy in a building are an intermediate indication of depletion of these natural resources. Life cycle energy analysis which determines the energy-related environmental impacts of buildings is therefore a more building specific and manageable quantitative approach. Since energy is the only measure used in determining environmental impact, a more detailed analysis over the lifetime is possible on which to base design decisions. Although it could be argued that energy alone does not indicate the total environmental impact, production and consumption of energy is a heavy burden on the environment in terms of depletion of resources and associated emissions, and forms a 'shorthand' or proxy for overall environmental impact.

Life cycle energy analysis

Life cycle energy analysis establishes the total energy attributable to a house over its lifetime. This is a quantitative method of analysis that is used to determine energy-related environmental impacts of buildings. Life cycle energy consists of both embodied and operating energy considered in concert. Energy associated with residential buildings is discussed in detail in the next chapter.

Embodied energy analysis

Embodied energy analysis is a scientific method for calculating the energy required to produce a particular product or a service, and in turn can be used to evaluate the environmental impact of production activity. However, many types of building material are produced in one country and sold and used in another. Hence the energy mix to be used for evaluation of embodied energy of such materials can become a complex issue. As an example, electricity in

Table 2.2 Comparison of embodied energy of selected building materials

Material	Baird and Chan (1983) (Input–output analysis)	Alcorn (1996) (Process analysis)	Alcorn (2003) (Hybrid analysis)
Cement, dry process	8.98 MJ/kg	7.8 MJ/kg	5.8 MJ/kg
Concrete, ready-mix	3,840 MJ/m³	2,350 MJ/m³	2,019 MJ/m³
Steel, recycled, reinforcing	59 MJ/kg	8.9 MJ/kg	8.6 MJ/kg
Timber, glulam	4,500 MJ/m³	2,530 MJ/m³	5,727 MJ/m³

New Zealand has a 64% hydro component (MED 2005: p. 120), whereas 77% of electricity in Australia is generated from coal (Uranium Information Centre 2006). It is possible to purchase the same product, for example paint, made in both countries, but the environmental impact of the paint will change with the source of the electricity used in its manufacture. The general convention in embodied energy analysis is not to include solar energy or human labour in such calculations.

Research has shown (Pears 1996: pp. 15–21) that there is no single correct method of building-embodied energy analysis. However, it should be noted that embodied energy coefficients calculated using different methods of analysis and different system boundaries vary considerably and are the usual cause of conflicting conclusions on the impact of the embodied energy of a building (see Table 2.2). Owing to the infinite number of processes involved, the use of the system boundary concept has become essential for embodied energy analysis as well. Many Australian models of embodied energy (Treloar 1996: pp. 51–58; Fay 1999) include energy attributable to third-order items such as banking and insurance. Therefore, these values tend to be larger than their equivalents in New Zealand. The results of the analysis depend not only on the system boundary but on the analysis method selected. The most common methods of analysis are statistical analysis, process analysis, input-output analysis and hybrid analysis. These are briefly outlined below.

Statistical analysis

This analysis relies on the published data on energy use by individual industries. The energy is then transferred to the proportions of products made by those industries. The method is very quick and easy provided that reliable data are available.

Process analysis

This is regarded as the most accurate and most common method (Alcorn 1996: p. 7) as it focuses on the energy requirements of a particular process. All processes outwith the defined system boundary are disregarded. Analysis begins with establishing the direct energy inputs to the production process under consideration and then traces back all the other contributory processes with their raw material requirements. Since all the direct energy requirements of all processes within the system are considered, this provides a reasonably accurate result. However, the effort required to account for all the tributary processes involved is not justifiable in most cases, while the method is also very time consuming owing to the complexity of an often increasing number of interconnected processes. The results are case specific and therefore are not representative of similar processes at other locations or times. Further, there may be processes for which data cannot be obtained owing to commercial sensitivity, lack of data collection or lack of data in an easily retrievable form.

Input–output analysis

This method uses national statistical information on inter-industry monetary flows, in terms of input–output tables. These provide information on economic contributions to and from a sector to the national economy. Using these economic flows to and from energy producing sectors and their total energy production values, it is possible to derive energy tariffs (MJ/$) for each sector in the economy. While the method is less time consuming and data are easily available, the results are also representative of other similar processes, so cannot be highly material specific. However, the method also captures all the minor tributory processes which are impossible to trace using process analysis. One of the main disadvantages of the method is that several dissimilar products with vastly different energy requirements may come under the same sector of the economy. For example, glass products, with a relatively high energy intensity (16–26 MJ/kg), are in the same category as concrete products, with relatively low energy intensity (0.9–2 MJ/kg). Further, there could be wide variation in the prices of products owing to market forces independent of energy use that is not represented in input–output tables.

Hybrid analysis

This method is a combination of the useful features of the process and input–output analysis methods. Analysis begins with process analysis of direct energy requirements and when the effort required to acquire data for process analysis outweighs the accuracy, the analysis reverts to the input–output method. Hence the method embraces both the completeness of input–output analysis and the reliability of the direct energy requirements of the process analysis method.

Embodied energy analysis in New Zealand

The first extensive research into the embodied energy of New Zealand house construction was carried out in 1983 by Baird and Chan (1983), using the input–output analysis method. The energy of human labour and the cost of environmental degradation were not included in their calculations. The lack of a standard method of assessing the contribution and proportion of the labour force and the lack of New Zealand data on environmental effects were the reasons for following the general convention of omitting these inputs. This should be borne in mind when energy coefficients of New Zealand building materials are compared with those of less industrialised countries with a higher labour input. In the current construction industry of New Zealand, which is largely factory based, site work consists mainly of assembling prefabricated components. According to Baird and Chan the direct energy requirement for site work and transport of materials was only 6% of the total gross energy requirement.

Baird and Chan's initial calculations, based on the New Zealand Department of Statistics' 130 industry input–output tables for 1971/72, when compared with the computations of another project (Carter et al. 1981), were found to be very low. This might be a result of the use of average energy prices for all the consuming sectors in this study. Therefore, Baird and Chan concluded that different energy prices paid by different sectors play an important role in calculation of energy intensities. Since their calculations were based on input–output tables for 1971/72, the published energy-intensity figures, which are sensitive to market forces, based on this method of analysis, were outdated even before being published in 1983. As has been suggested by other researchers (Bullard et al. 1976; Alcorn 1996), the use of input–output tables alone for energy analysis of this kind is unsuitable. Owing to process and technological advancements and efficiency in use of energy, embodied energy coefficient figures for New Zealand have since been updated three times (Alcorn 1996; Alcorn and Wood 1998; Alcorn 2003). New information that has become available since 1983 from industry, academic and overseas sources could have improved the accuracy of these energy calculations. Calculated energy intensities are for the first use of materials. Hence, neither the energy for recovery of material for recycling, reuse or reprocessing, which is attributable to second use, nor the energy for demolition or disposal of materials, is included.

Operating energy analysis

Operating energy analysis is fairly straightforward as it is either measured or monitored. For the residential sector this can be done either by dividing the national domestic sector energy use by the population or the number of houses, or by monitoring a series of representative houses.

Operating energy analysis in New Zealand

In previous energy studies (Baird and Chan 1983; Breuer 1988; Honey and Buchanan 1992; Jaques 1996) of New Zealand residential constructions, the operating energy for appliances and equipment was not included. Even though the operating energy requirement is heavily dependent on user behaviour, based on average consumption figures available Baird and Chan established that the ratio of construction-embodied energy to operating energy varies from 1:7 to 1:11 depending on the construction type used. In a similar study by the Building Research Association of New Zealand (BRANZ), energy used for heating, lighting and hot water services in NZ houses over a lifetime of 50 years was found to be four times the construction energy and 23 times the total maintenance energy requirement (Jaques 1996: p. 10). This is a good example of LCA revealing which the major envrionmental impacts of a product are, in this instance houses.

Conclusions

Life cycle analysis is a quantitative method used to determine the environmental implications of resource use. Owing to inherent weaknesses in the analytic method and the relatively longer lifetime of buildings, compared with most other products to which LCA is applied the results do not provide absolute values for data; however, it is still a useful tool for quantification of the environmental impact. Life cycle energy analysis is a more building specific and manageable method for impact analysis based mainly on building-related energy use.

Embodied energy analysis is also problematic owing to the use of vastly different system boundaries and the lack of agreement on the methodology to be used. As a result, the energy coefficients calculated using different methods and system boundaries vary widely. Embodied energy values for New Zealand construction materials and elements have been established and updated. Some of the Australian embodied energy coefficients are considerably higher than their New Zealand equivalents. Thus designers have to be aware of the country of origin of products in undertaking an LCA. Operating energy analysis is well established and straightforward. However, previous operating energy calculations of NZ houses have not included energy use for appliances and equipment.

The next chapter examines the life cycle energy of houses.

References

Adalberth, K. (1997) Energy use during life cycle of buildings: A method. *Building and Environment* 32(4): 317–320.

Alcorn, A. (1996) *Embodied Energy Coefficients of Building Materials.* Centre for Building Performance Research, Victoria University of Wellington.

Alcorn, A. (2003) *Embodied Energy and CO_2 Coefficients for NZ Building Materials.* Centre for Building Performance Research, Victoria University of Wellington.

Alcorn, A. and Wood, P. (1998) *New Zealand Building Materials Embodied Energy Coefficients Database, Volume II – Coefficients.* Centre for Building Performance Research, Victoria University of Wellington.

Alcorn, J. A. and Haslam, P. J. (1996) The Embodied Energy of a Standard House – Then and Now. In: *Proceedings of the Embodied Energy: The Current State of Play Seminar* (G. Treloar, R. Fay and S. Tucker, eds), pp. 133–140. Deakin University, Australia.

Antonsson, A–B. and Carlsson, H. (1995) The basis for a method to integrate work environment in life cycle assessments. *Journal of Cleaner Production* 3(4): 215–220.

Baird, G. and Chan, S. A. (1983) *Energy Cost of Houses and Light Construction Buildings* (Report No.76), New Zealand Energy Research and Development Committee, University of Auckland.

Barnthouse, L., Fava, J., Humphreys, K., et al. (1998) *Life Cycle Impact Assessment: The State-of-the-Art*, 2nd edn. SETAC Press.

Bekker, P. C. F. (1982) A life-cycle approach in building. *Building and Environment* 17(1): 55–61.

Breuer, D. (1988) *Energy and Comfort Performance Monitoring of Passive Solar, Energy Efficient New Zealand Residences* (Report No. 172), New Zealand Energy Research and Development Committee, University of Auckland.

Bullard, C. W., Penner, P. S. and Pilati, D. A. (1976) *Net Energy Analysis: Handbook for Combining Process and Input-Output Analysis.* Energy Research Group, Center for Advanced Computation, University of Illinois at Urbana-Champaign.

Carter, A. J, Peet, N. J. and Baines, J. T. (1981) *Direct and Indirect Energy Requirements of the New Zealand Economy: An Energy Analysis of the 1971–72 Inter-industry Survey.* New Zealand Energy Research and Development Committee, University of Auckland.

Chevalier, J. L. and LeTeno, J. F. (1996) Requirements for an LCA-based model for the evaluation of the environmental quality of building products. *Building and Environment* 31(5): 487–491.

Erlandsson, M. and Levin, P. (2005) Environmental assessment of rebuilding and possible performance improvements effect on a national scale. *Building and Environment* 40(11): 1459–1471.

Fay, M. R. (1999) *Comparative Life Cycle Energy Studies of Typical Australian Suburban Dwellings.* PhD Thesis, Deakin University, Australia.

Hobbs, S. et al. (1996) Sustainable Use of Construction Materials. *Proceedings of Sustainable Use of Construction Materials.* Building Research Establishment, UK.

Honey, B. G. and Buchanan, A. H. (1992) *Environmental Impacts of the New Zealand Building Industry* (Research Report 92/2). Department of Civil Engineering, University of Canterbury, New Zealand.

ISO 14040 (1997) *Environmental Management – Life Cycle Assessment: Principles and Framework*, 1st edn. International Standardization Organization.

Jaques, R. (1996) Energy Efficiency Building Standards Project: Review of Embodied Energy. In: *Proceedings of the Embodied Energy: The Current State of Play Seminar* (G. Treloar, R. Fay and S. Tucker, eds), pp. 7–14. Deakin University, Australia.

Jaques, R. (1998) Cradle to the Grave – LCA Tools for Sustainable Development. *32nd Annual Conference of the Australia and New Zealand Architectural Science Association, Wellington, New Zealand*, 15–17 July 1998 (reprint *BRANZ Conference Paper No. 47, 1998*).

Johnstone, I. M. (2001) Energy and mass flows of housing: Estimating mortality. *Building and Environment* 36(1): 43–51.

Jönsson, Å., Tillman, A. M. and Svensson, T. (1996) Life cycle assessment of flooring materials: Case study. *Building and Environment* 32(3): 245–255.

Jönsson, Å. (2000) Is it feasible to address indoor climate issues in LCA? *Environmental Impact Assessment Review* 20(2): 241–259.

Krozer, J. and Vis, J. C. (1998) How to get LCA in the right direction? *Journal of Cleaner Production* 6(1): 53–61.

Li, Z. (2006) A new life cycle impact assessment approach for buildings. *Building and Environment* 41(10): 1414–1422.

Lindeijer, E. (1996) Normalization and Valuation. In: *Towards a Methodology for Life Cycle Impact Assessment* (Udo de Haes, ed.). SETAC.

Ministry of Economic Development (2005) *New Zealand Energy Data File – July 2005* (compiled by Hien, D. T. Dang). Energy Modelling and Statistics Unit, Ministry of Economic Development.

Pears, A. (1996) Practical and Policy Issues in Analysis of Embodied Energy and its Application. In: *Proceedings of the Embodied Energy: The Current State of Play Seminar* (G. Treloar, R. Fay and S. Tucker, eds), pp. 15–22. Deakin University, Australia.

REGENER Project (1997) European methodology for the evaluation of environmental impact of buildings, Regener Project final report, Available at: http://www-cenerg.ensmp.fr/francais/themes/cycle/html/11.html [accessed 14 March 2006].

Treloar, G. J. (1996) A Complete Model of Embodied Energy 'Pathways' for Residential Buildings. In: *Proceedings of the Embodied Energy: The Current State of Play Seminar* (G. Treloar, R. Fay and S. Tucker, eds), pp. 51–58. Deakin University, Australia.

Uranium Information Centre (2006) *Australia's electricity*. Available at: http://www.uic.com.au/nip37.htm [accessed 3 May 2006].

Verschoor, A. H. and Reijnders. L. (1999) The use of life cycle methods by seven major companies. *Journal of Cleaner Production* 7(5): 375–382.

Wright, J. and Baines, J. (1986) *Supply Curves of Conserved Energy: The Potential for Conservation in New Zealand's Houses*. Centre for Resource Management, University of Canterbury, New Zealand.

Wu, X., Zhang, Z. and Chen, Y. (2005) Study of the environmental impacts based on the 'green tax' – applied to several types of building materials. *Building and Environment* 40(2): 227–237.

Zhang, Z., Wu, X., Yang, X. and Zhu, Y. (2006) BEPAS – a life cycle building environmental performance assessment model. *Building and Environment* 41(5): 669–675.

3 Life cycle energy of houses

Energy is sought for the work or the services it provides in the industrial, commercial, transport and household sectors and for many other activities. Flow of energy is based on the two basic laws of thermodynamics, which state that:

1. energy cannot be created or destroyed, and therefore the sum of total energy in the universe is fixed;
2. energy moves from a concentrated state to a more dispersed state, and therefore the amount of ordered energy in the universe is decreasing.

As a result, the higher the order of energy used for an activity the greater will be the environmental impact of that activity because of the reduction in availability of high-order energy. Driving the car a short distance to collect the daily newspaper has a considerable environmental impact as the same task could be done with the lower-order energy involved in walking.

Energy sources

Based on published records, historically, energy services have made up around 2.7% of the gross domestic product (GDP) in New Zealand (MED 2001: p. 7). Fossil fuels, which are the remains of long-dead forests and marine life carbohydrates that have been transformed into hydrocarbons, form about 69% of New Zealand's energy supply. Hydroelectricity and geothermal steam make up a major part of the balance. Wind, biogas, industrial waste and wood, and solar water heating are the other minor energy sources (MED 2005: p. 9).

Primary energy is the total amount of energy available for use. While some of this primary energy is used directly, most of it is converted from its initial state to a more convenient state known as delivered energy for day-to-day use. Primary energy is the energy as it is first obtained from natural sources such as coal as it is mined, hydro as it is used for electricity generation, etc. When this conversion from primary energy to delivered energy involves heat, large amounts of energy are lost. Only 15% of the primary energy in geothermal steam and waste is actually converted to electricity while the figure is 35% for gas, 33% for coal and, 100% for hydro and wind (MED 2005: p. 15). Therefore, total energy consumption is considerably greater than end use may suggest.

Energy is measured in joules (J), megajoules (MJ) or petajoules (PJ). Power is the rate at which energy is converted from one form to another. The watt (W) is the measure of power. Therefore, one watt of electricity is the conversion of one joule in a second.

Energy sources can be broadly divided into two major categories, renewable and non-renewable.

Renewable energy

Renewable resources are defined as those resources that are created or produced as fast as they are consumed, with nothing being depleted (Barnett and Browning 1995: p. 101). In 2004, 31% of the primary energy sources in New Zealand were renewable energy sources compared to 1.4% in the UK (DTI[1] 2006). Hydroelectricity and geothermal energy, accounting for 12.7% and 11.0% of primary energy respectively, are the two main sources of energy in New Zealand (MED 2005: p. 9). Geothermal energy is mainly used in the form of electricity, although it also provides hot water in some areas.

Other renewable energy sources available in New Zealand are, wind, direct sunlight, biofuels, industrial waste and wood. Being a relatively narrow set of islands, New Zealand is windier than most countries. However, wind energy generation is only now becoming more common. Wood is still a major fuel for domestic heating in rural areas of New Zealand (Statistics New Zealand 2002: p. 14) and in some urban areas such as Christchurch where its use has led to an air pollution problem. Recent research (Isaacs et al. 2005: p. 51) suggests that wood use currently accounts for around 15–20% of domestic energy consumption. Various regulations that come under 'National Environmental Standards for Air Quality' (AS/NZS3580.9.10:2006), which include a Standard for wood burners aimed at controlling the air pollution problem, and similar legislation such as Environment Canterbury's Natural Resources Regional Plan[2], could lead to higher use of other energy sources when people shift away from wood fires.

Non-renewable energy

Non-renewable resources are defined as those resources of which the supply is limited or which can be depleted to a degree rendering recovery to be too costly (Pilatowicz 1995: p. 24). Oil, natural gas and coal are the main non-renewable energy sources used in New Zealand. Together they constitute 69% of primary energy. In 2004, the primary energy supply of New Zealand constituted: oil (36%); gas (21%); coal (12%); hydro (13%); geothermal (11%); and other renewables (7%) (MED 2005: p. 9).

Energy use

New Zealand uses about 126 GJ[3] of primary energy per person each year (MED 2005: p. 7). About one third of this is lost in extraction and conversion to delivered (consumer) energy, and the balance is used as electricity and fuels. Although only a part of this energy is used directly, the balance is used indirectly as the energy embodied in the products being used.

The energy intensity of a country is the amount of energy used to produce a unit of GDP. (Energy intensity when measured in *toe* – tonnes of oil equivalent, represents energy produced by burning one metric ton of crude oil, which is roughly equivalent to 41.868 GJ or 11.64 MWh.) Energy intensity and associated CO_2 emissions for selected regions and countries in the year 2003 are as shown in Table 3.1. However, it could be argued that this is a crude measure of energy intensity, since the type of energy used is not taken into account. The primary energy of renewable sources does not matter to the same extent as the primary energy of non-renewable sources, and 31% of New Zealand primary energy comes from renewable sources. Although 'lost' primary energy from renewable sources is harmless, 'lost' primary energy from non-renewable sources causes significant environmental damage, for example through climate change. Therefore if the primary energy from renewable sources is deducted the energy intensity of New Zealand is only 207 toe/million US$. This factor is accounted for in the carbon intensity figures that accompany the energy intensity figures shown in Table 3.1.

Although the energy intensity of Australia is lower than that of New Zealand, the CO_2 emissions per US$ of GDP are higher as a result of the higher use of coal as an energy source. Moreover, New Zealand uses energy to produce primary products, such as aluminium and timber for the building industry. Although Japan has reduced the energy intensity of production in recent times, as argued by Goodland and Daly (1996: p. 1013), this has only been achieved by de-linking energy and production by importing all timber requirements and smelting most of the aluminium required overseas.

However, the above comparison using GDP based on the market exchange rate does not take into consideration the difference in living standards and consumption patterns in various countries. A dollar spent in two different countries could

Table 3.1 Energy intensity and CO_2 emissions of selected countries and regions, 2003

Region/country	TPES[†]/GDP[#] (toe/US$)	CO$_2$/GDP[#] (kg/US$)
Africa	870	1.19
Asia	720	1.38
Middle East	660	1.63
Latin America	320	0.59
OECD	200	0.48
World	**320**	**0.75**
Russia	2,090	4.98
China	1,020	2.70
India	1,020	1.93
Turkey	380	0.96
Canada	340	0.72
New Zealand	**300**	**0.56**
Mexico	270	0.63
Australia	260	0.81
USA	220	0.55
France	200	0.29
Germany	180	0.45
UK	150	0.35
Denmark	130	0.34
Norway	130	0.20
Japan	110	0.25

[†] Total primary energy supply (in tonnes of oil equivalent)
[#] Gross domestic product (in US$ of year 2000 value)
Based on: International Energy Agency (2005) *Key World Energy Statistics*. Available at: http://www.oecd.org/topicstatsportal/0,2647,en_2825_495616_1_1_1_1_1,00.html (accessed 27 March 2006)

have vastly different worth depending on the economic situation. Table 3.2 shows energy intensity and CO_2 emissions for the same regions and countries based on purchasing power parity (values adjusted to address the variations in economies).

In 2004, 12% of total consumer energy use in New Zealand was in the residential sector (MED 2005: p. 13), while in the UK 31% of the total consumer energy use was in the residential sector (DTI 2002: p. 11). Residential sector energy use according to fuel type is shown in Table 3.3. In the year 2004, average residential

Table 3.2 Energy intensity and CO_2 emissions for selected countries and regions, 2003 (based on purchasing power parity)

Region/country	TPES[†]/GDP (PPP)[*] (toe/US$ PPP)	CO₂/GDP (PPP)[*] (kg/US$ PPP)
Middle East	380	0.95
Africa	300	0.40
Asia	190	0.37
OECD	190	0.45
Latin America	160	0.29
World	**210**	**0.51**
Russia	510	1.22
Canada	280	0.60
China	230	0.61
USA	220	0.55
Australia	200	0.61
New Zealand	**200**	**0.37**
India	190	0.36
Germany	170	0.41
Mexico	170	0.41
France	170	0.24
Turkey	160	0.42
Japan	150	0.35
UK	140	0.34
Norway	140	0.21
Denmark	130	0.36

[†] Total primary energy supply (in tonnes of oil equivalent)
[*] Gross domestic product based on purchasing power parity (in US$ of year 2000 value)
Based on: International Energy Agency (2005) *Key World Energy Statistics*. Available at: http://www.oecd.org/topicstatsportal/0,2647,en_2825_495616_1_1_1_1_1,00.html (accessed 27 March 2006)

sector electricity consumption in New Zealand was 7.82 MWh per household or 28 GJ per household (MED 2005: p. 133).

However, the percentage contribution by renewable sources in New Zealand according to Table 3.3 is lower compared with the estimates arrived at on the basis of the recent monitoring of houses in the HEEP study (Isaacs et al. 2005)

Table 3.3 Consumer energy use in the residential sector by fuel type, 2004

Fuel type	New Zealand	UK
Electricity	73%	9%
Gas	12%	66%
Oil	4%	5%
Renewables	10%[†]	2%
Solid fuels	1%[#]	18%

[†] geothermal, solar and wood
[#] mainly coal
Based on: Department of Trade and Industry (2006) *Energy Consumption Tables*. Available at: http://www.dti.gov.uk/energy/inform/energy_consumption (accessed 30 March 2006) and Ministry of Economic Development (2005) *New Zealand Energy Data File – July 2005* (compiled by Hien, D. T. Dang), Energy Modelling and Statistics Unit, Energy and Resources division, Ministry of Economic Development)

discussed earlier (see Renewable energy above). Further discussion of HEEP results and methodology is given later in the text.

Energy use in residential buildings

The energy associated with a building depends on:

- physical characteristics – the construction type and workmanship; the location, orientation, size and shape of the building; fenestration; etc.;
- equipment related issues – the type and number of appliances and lights; and
- occupant related issues – family: size, composition, age structure, income level, comfort expectations, etc.

Factors such as ownership and occupant behaviour may also affect the energy associated with a building. The average floor area of an NZ house has been steadily rising over the years. The average floor area of a new house increased by 25% from 146 m^2 in 1970 to 194 m^2 in 2000 (CfHR[4] 2004: p. 7). Larger houses are associated with higher levels of energy both in construction and operation, thereby leading to an increase in the energy used in the residential sector.

Energy attributable to residential buildings (and to all types of building) is of three types:

1. embodied energy
2. operating energy
3. life cycle energy.

These are discussed in the following sections.

Embodied energy

Embodied energy is normally defined as not only the energy directly consumed by a process but the energy indirectly used to produce goods and services associated with the process (Baird and Chan 1983: p. 4). It is also possible to consider the energy involved in things like insurance companies that might insure a manufacturer of materials. Therefore, the energy embodied in houses (and other buildings) consists of:

- direct energy – energy used at the site to assemble the materials and to transport the materials to site;
- indirect energy – energy used in mining raw materials, manufacture of materials, tools and machinery, insurance, banking and other related services, etc.

The sum of the direct and indirect embodied energy is known as GER or the gross energy requirement. Embodied energy could also be expressed as an energy coefficient or energy intensity and is denoted by MJ/kg for materials, MJ/m^2 for constructions (walls, floors, windows, etc.) or whole buildings, and MJ/m^3 for both materials and whole buildings.

Operating energy

Operating energy is the energy required for space conditioning (heating and cooling), hot water, cooking and lighting together with the energy for operation of appliances and equipment. Cooling is not a general practice in the domestic sector in New Zealand at present (see later Figure 3.1), while 7% of the total energy use in the average US house in 2001 was for cooling.

Space heating

The Household Energy End-use Project, commonly known as HEEP, is a long-term research project, which commenced in 1995 with the aim of determining energy use in residential buildings representative of current New Zealand practices. The information collected by the study includes:

- energy – total energy and end uses in an individual house;
- temperature – up to 10 points within the house and one point outside;
- climate – solar radiation, wind speed, rainfall, temperature;
- survey – building thermal performance; number, type and use of appliances; household demographic characteristics; occupant attitudes and behaviour (Stoecklein et al. 1998b).

Design recommendations geared to the provision of thermal comfort are generally given in terms of the internal room temperatures (air temperatures). New Zealand is a temperate country with only a small range of extreme weather. Although comfort conditions of New Zealanders have not been investigated, anecdotal comments suggest that 18°C would be an acceptable indoor temperature in winter (Isaacs 1998: p. 3). Energy used for space heating is dependent *inter alia* on comfort expectations, affordability, i.e. income level and energy prices, external temperature, ventilation rates, heat loss from the building fabric and user behaviour.

The space heating energy requirement, normally referred to as the auxiliary heating demand, depends on the difference between the gross heat loss from the building and various free casual heat gains such as solar gains, and internal gains – the heat emitted by the occupants and waste heat from appliances and lighting. Owing to the effect of free heat gains, the amount of energy purchased is much less than the actual energy required for space heating. However, heat gains cannot generally all be utilised to reduce the heating energy requirement as they may occur when the occupants demand no specific temperature level. A fraction of such gains can be stored in the thermal mass of the building (thermal mass is dense building materials which can absorb and retain large amounts of heat) and be released later. While free heat contributes to maintaining thermal comfort during the cooler months, it is wasted during the summer months or could even lead to overheating.

Typical heat gains from occupants range from 70 W for an adult when sleeping to 352 W when carrying out heavy work over a period of eight hours (Vale and Vale 1991: p. 43). Previous research by the Building Research Association of New Zealand (BRANZ) has established that for an average domestic building, body heat amounts to about 6 kWh/day and waste heat from appliances about 10 kWh/day (Trethowen and Bassett 1979: p. 125). According to Isaacs (1998: p. 4), a house with the benefit of a few occupants and their activities will provide additional warmth of about 3°C above the external temperature. Therefore, as per Table 3.4, in Auckland, where 34% of the population of New Zealand live, a moderately well designed house should be able to provide comfortable conditions, or very close to them, without any purchased heating.

Table 3.4 Climate indicators for main centres in New Zealand, 1998[#]

Location	Avg. temp. (°C)	Max. temp. (°C)	Min. temp. (°C)	Avg. sunshine (h)	Avg. rainfall (mm)
Auckland	15.3	32.4	−0.1	2,102	1,185
Wellington	12.5	31.1	−1.9	2,019	1,240
Christchurch	11.6	41.6	−7.1	1,974	666
Dunedin	10.8	34.4	−8.2	1,676	938

[#] Values are annual averages
Based on: Rawlinson & Co. (ed.) *Rawlinsons: New Zealand Construction Handbook*. Rawlinsons New Zealand Constructions Handbook Ltd., 1998, p. 614

Energy use for space heating, which is partly dependent on the external temperature, varies widely over New Zealand. According to the HEEP study in the Auckland region space heating (17%) and hot water (28%) together are estimated to represent 45% of the total residential electricity consumption (Isaacs 2004: p. 7). However, this is not the total energy consumption, because much residential heating comes from wood. In an average New Zealand house 30% of the total energy use (using all fuel types) is estimated to be for space heating while a further 29% is estimated to be for water heating (Isaacs 2004: p. 8). Annual electricity consumption by end use in an Auckland house and average domestic energy use by end use based on HEEP estimates are shown in Figure 3.1.

In the UK, 62% of total domestic energy use in 2001 was for space heating (DTI 2006). According to a residential energy consumption survey undertaken in the USA, energy consumption in the average US house in 2001 was 100 GJ, of which 46% was used for space heating (Energy Information Administration 2004) (see Figure 3.2).

In comparison with the UK and USA, heating energy use in Auckland as a percentage of the total is very low. Although this is partly due to the mild weather characteristics of New Zealand, which lead to a shorter heating season, the lower comfort temperature expectations of New Zealanders also contribute to this. The indoor climate maintained within residential buildings determines the amount of energy consumed for space heating as well as the comfort and health of the occupants. Many houses in New Zealand are insufficiently heated and the temperatures in some spaces fall well below the WHO (World Health Organization) recommended levels for indoor temperatures (Pollard et al. 1998). Annual residential energy consumption in a house in NZ, the UK and the USA are as shown in Table 3.5.

Buildings play an important role in supporting the state of health. WHO recommend that internal temperature be maintained between 18 and 24°C with air

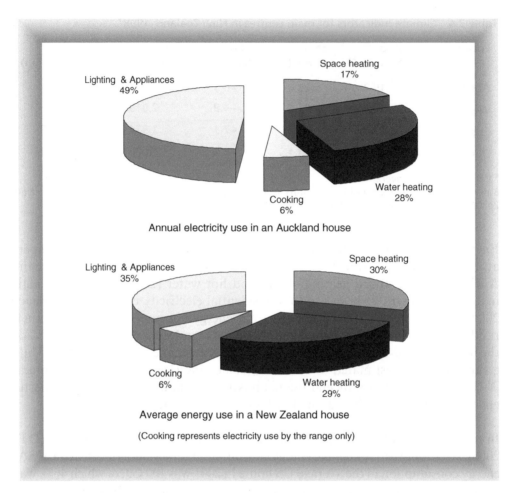

Lighting & Appliances
49%

Space heating
17%

Cooking
6%

Water heating
28%

Annual electricity use in an Auckland house

Lighting & Appliances
35%

Space heating
30%

Cooking
6%

Water heating
29%

Average energy use in a New Zealand house

(Cooking represents electricity use by the range only)

Fig. 3.1 Comparison of electricity use in an Auckland house and total energy use in an average New Zealand house by end use.
Based on: Isaacs, N. (2004) Supply Requires Demand – Where does all of New Zealand's energy go?, *Royal Society of New Zealand Conference*, Christchurch, 18 November 2004, pp. 7–8. (reprint *BRANZ Conference Paper No. 110, 2004*).

speeds of less than 0.2 m/s. This is to be coupled with 50% relative humidity (Bell et al. 1996: p. 74). Lower internal temperatures also lead to buildings suffering from dampness caused mainly by the condensation of water vapour in the air. In addition to the water vapour naturally present in the air, people and their activities add a considerable amount thereto. The amount of moisture added by various sources to the air inside a house according to the British Standards quoted by Isaacs (1999: p. 4) is as per Table 3.6.

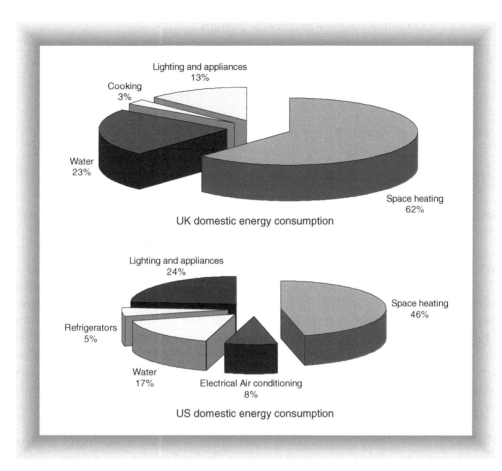

Fig. 3.2 Comparison of domestic energy consumption by end use in the UK and USA (2001). Based on: Department of Trade and Industry (2006) *Table 3.6: Domestic Energy Consumption by end use 1970 to 2003* Available at: http://www.dti.gov.uk/energy/inform/energy_consumption/ecuk3_6.xls (accessed 30 March 2006) and Energy Information Administration (2004). *Residential Energy Consumption Surveys: 2001 consumption and expenditure tables.* Available at: http://www.eia.doe.gov/emeu/recs/contents.html (accessed 31 March 2006).

Table 3.5 Comparison of residential energy consumption in 2001

Country	Energy use (GJ/house)	Data sources
New Zealand	46	Ministry of Economic Development 2002 and Statistics New Zealand 2002
UK	81	http://www.dti.gov.uk/energy/inform/energy_consumption
USA	100	Energy Information Administration 2004

Table 3.6   Additions of moisture to domestic interiors

Source	Rate
Unflued LPG or natural gas heater	0.3–0.6 kg/h
Cooking	2 kg/day electric cooker 3 kg/day gas cooker
Bathing (showers, baths and hand washing)	200 g/day/person
Clothes drying (unvented)	1–1.5 kg/person/day
Humans	40 g/h sleeping 55 g/h active

Source: BS 5925:1991 *Code of Practice for Ventilation Principles and Designing for Natural Ventilation*. London: British Standards Institution, 1991

When the amount of water vapour in air exceeds the amount that it can hold at that temperature, and if there is no ventilation to exchange moisture-laden indoor air for external air with lower humidity, the additional vapour condenses out on the nearest cold surface in the form of dew – most likely on windows. During cold damp nights, clothes, room linings, framing and furnishings absorb the moisture and release it during the warm day. Therefore, a correct level of ventilation is essential for good health. BRANZ research has found that while older houses have excess ventilation, newer houses are more airtight and likely to be short of ventilation if windows are not opened (Isaacs 1999: p. 6).

Electricity, gas and solid fuels (coal and wood) are the main sources of energy used for space heating. In 2001, 72% of houses in New Zealand used electricity for heating. Although the use of wood is as high as 76% of houses in some areas, it is very low in both Auckland and Wellington. Use of gas as a fuel has increased over the years, owing to increased use of gas appliances for space heating (Statistics New Zealand 2002). The capital cost of portable LPG heaters is higher than for electric heaters. Furthermore they are expensive to run (Isaacs 1998), contribute unwanted moisture to the internal spaces and can be dangerous if not properly operated. Without proper ventilation the use of LPG heaters in smaller spaces such as bedrooms could be fatal. However, because of their rapid heating capability and the controllability of operating costs they are very popular among low-income households.

Affordability is closely associated with space heating energy use. Affordability depends on both income level and the cost of energy. In 2001, 70% of households in New Zealand had an income in excess of NZ$ 25,000, while 50% had an income of over NZ$ 50,000 (Statistics New Zealand 2002: p. 18). As indicated by Table 3.7, at present the price of electricity in New Zealand is low compared with most other OECD countries. Almost all of the cheap and easy hydroelectric sites

Table 3.7   Residential electricity prices for selected countries, 2002

Country	Price of electricity for households	
	US$/kWh	US$PPP/kWh[#]
Norway	0.045	0.038
Australia[*]	0.062	0.084
USA[*]	0.085	0.087
New Zealand	**0.071**	**0.105**
UK	0.105	0.111
France	0.105	0.122
Mexico	0.092	0.134
Germany	0.136	0.146
Japan	0.174	0.150
Denmark	0.209	0.190
Turkey	0.099	0.243

[*] Price excluding tax
[#] Price per kilowatt hour in US$ adjusted based on purchasing power parity
Based on: International Energy Agency (2004) *Electricity Information*. IEA Statistics, pp. 1.67, 1.68

have already been developed. When new and more costly plants are installed, it is expected that electricity prices will start to rise.

The net space heating energy requirement of a house depends on the heat gains and losses through the fabric. Houses gain heat from the sun, occupants and waste heat from appliances. Heat losses occur via five routes, i.e. roof, walls, floors, windows and replacement of warm air by cold air, termed infiltration and ventilation. Heat loss from the building fabric depends *inter alia* on thermal insulation and effects of ventilation. For the best performance of the thermal fabric, all of these aspects should be considered in the design process.

According to a survey carried out by BRANZ in 1985, 15% of total fabric heat loss of a house insulated to the New Zealand Standard occurs through infiltration. Bassett (1985) found a correlation between leakiness and complexity of the envelope. A simple box shape is likely to have a low infiltration rate because it has fewer joints to be sealed. With the realisation that heat retention is more important than collection, glazing areas have been limited in the new building code (NZS 4218:1996) to 30% of the external wall surface area. According to the same standards, insulation requirements for housing in the cool zone of New Zealand have to satisfy higher performance standards. This cool zone comprises South Island and the central plateau of North Island.

The benefit of increased levels of insulation may be a reduction in the use of energy or an increase in comfort, or a combination thereof. Trethowen (1979) argues that it is reasonable to assume that one third of the benefit of insulation is normally taken as improved comfort. The HEEP study has recorded that, on average, houses built since the minimum levels of insulation became mandatory in 1978 are 1°C warmer while the total energy intensity of NZ houses also decreased by 8% during the period 1990 to 2004 (Isaacs 2004). Similarly in the UK, the average UK house internal temperatures increased from 13°C in 1970 to 18°C in 2000, while the number of houses with central heating also increased from less than a third in 1970 to 89% in 2000. However, the residential energy intensity decreased by 6% between 1970 and 2000 (DTI 2002). Apart from reduced energy use and higher levels of thermal comfort, increased levels of insulation will have other impacts by reducing the potential for mould growth, increasing the durability of the structure and affecting humidity in the internal spaces. However, it should also be noted that theoretical performance of higher levels of insulation might not be achieved owing to construction practices, since, if insulation is not properly installed, the gaps in its continuity will lower its predicted effectiveness.

Water heating

A large percentage of the electricity used in a typical NZ house is for water heating (see Figure 3.1). A regular 1.5 kW of energy use throughout the day attributable to the demand for hot water was identified by the HEEP study (Isaacs 1997: p. 10). Although electricity use rises in the winter part of the year owing to the use of energy for space heating, this rise is partly attributable to the hot water system as well. Energy is consumed for water heating to maintain the storage temperature (due to standing losses) and to replace the hot water that has been used (consumed energy). In winter, cold water entering the hot water cylinder will be at a lower temperature than in summer.

A considerable amount of heat is lost through the cylinder wall and the distribution pipes. The higher the difference between the temperature of the hot water and the surrounding air, the greater the losses. According to the analysis by the HEEP study, the standing losses from the typical hot water system have been calculated to be in the range of 4.1 kWh/day (for gas storage) to 2.6 kWh/day (for electric storage), depending on the system. This constituted about 27–34% of the total energy consumption (Isaacs 2004: p. 10).

According to Parker and Tucker (1985), cited by Wright and Baines (1986), an average household uses around 200 litres of hot water daily. Hendtlass (1981) found that the amount of water used in a household depends on the composition of the household. Households with younger children and teenagers use a comparatively higher amount of hot water. The present New Zealand population consists of an aged population with 28% of people being in the age group 60–74 years[5] therefore, a reduction in the amount of hot water used in an average New Zealand household could be expected.

The HEEP study found that the hot water cylinder thermostat is often inaccurate and supplies water at unsafe temperatures. For 43% of the houses monitored, the water temperature was above 60°C while it was 70°C for another 13% of houses (Isaacs et al. 2004: pp. 1–145). The New Zealand Building Code Clause G12 Acceptable Solutions recommends a maximum delivery temperature of 45°C for the safety of younger children and the elderly and 55°C for all others. However, the cylinder temperature has to be maintained at 60°C to avoid the risk of the growth of *Legionella* bacteria. (Previous research has found *Legionella* in around 10% of houses, although live bacteria were not present in cylinders.) However, only 24% of households use a larger (180 litre) cylinder, and a lower set temperature may lead to some households with smaller cylinders (135 litres) running out of hot water regularly. Since a lower surface-to-volume ratio lessens relative losses, the use of larger cylinders is encouraged. According to the HEEP study, mean temperature was 61°C for smaller cylinders and 58°C for larger types (ibid.).

The main contributory factors to the high use of energy for water heating have been identified as the traditional use of hot water for laundry, and the use of waste heat from hot water cylinders for airing clothes and linen in airing cupboards. Although installation of high thermal insulation grade 'A' hot water cylinders as replacements could save around 0.1% of total New Zealand electricity use (if temperature is set to 55°C), space limitation for these large cylinders is a drawback (Harris et al. 1993).

Harris et al. (1993: p. 26) have suggested that domestic electricity use could be reduced by up to 13% with the simple measures listed below.

- reducing cylinder temperature to 55°C (6% saving);
- use of low-flow shower heads, kitchen and bathroom tap aerators, and cylinder wraps (5% saving);
- cold water clothes washing (2% saving).

However, it is not clear whether the role played by internal heat gains in providing a house with a significant amount of heat has been taken into account in the above suggestions. If it has not, the reduction in losses from the hot water system through reduction in cylinder temperature could be offset by an increase in the use of electricity for space heating at times to make up for the lost heat gains. Reducing the cylinder temperature below the 60°C mark, which is required by the Building Code to avoid the risk of *Legionella*, could lead to health problems.

Cooking

Cooking and eating habits vary among different ethnic groups. Over the past 10–15 years, owing to social and lifestyle changes, these habits have changed globally, considerably reducing the time and human energy spent on day-to-day cooking, with the introduction of various appliances in the process. Even though there has been a steady decrease in energy used by the kitchen range (from 13% in 1971 to 7% in 2000), the number of other appliances used in the actual cooking process has been rising. In the UK during the period 1970 to 2000, energy use in cooking was reduced by 16% (DTI 2002: p. 23).

The conventional oven has largely been replaced by the fan-assisted oven, leading to an approximate 25% reduction in energy use (Wright and Baines 1986: p. 55). With more women entering the paid workforce, 'heat and eat' has become more common. Ownership of a microwave oven has shown a steady rise. The substitution of the microwave oven for the conventional oven is estimated at 70 to 100% for all oven use (Wright and Baines 1986). In the year 2000, just over 70% of households in the UK owned a microwave oven (DTI 2002: p. 27).

Energy consumption relating to cooking also depends on eating habits. Access to and increased frequency of using takeaway and fast food services could lead to a reduction in energy use in the home, although this would not affect overall levels but merely shift energy use from the domestic arena to the commercial sector.

Lighting

Different people have different lighting energy requirements. Light quality, quantity and colour are perceived differently by people owing to varied personal visual characteristics and, hence, preferences. Studies have shown that an average 60-year-old person needs a lighting level 10% higher than that for a 20-year-old (Pilatowicz 1995: p. 57). There is an upward trend in the use of lighting. This is associated with special concerns such as security, safety and accident prevention, comfort, and increased use of lighting for special effects – location lighting, spotlights, etc. There is an increased use of outside lights for security. The HEEP study found that electricity use for lighting is dependent on income level (the higher the income the higher the use of energy) and the time people retire to bed (Isaacs 1997). Of the total of energy usage in a New Zealand house, 15% is for lighting (Isaacs 2004: p. 7). The highest contributors to the peak power demands that exist from 7.30 a.m. to 12:00 noon and 6.00 p.m. to 11.00 p.m. in New Zealand have been identified as domestic hot water and space heating, and lighting. Contribution to peak demand due to lighting is estimated to be 200 W per house (Isaacs 2004: p. 11). Energy consumption for lighting in the UK house increased by 13% between 1990 and 2000 owing to a shift towards multi-source lighting (use of wall and table lamps with multi-ceiling lights).

The main energy improvement in lighting consists in replacing incandescent lamps with compact fluorescent lamps (CFLs). 90% of the energy used by a standard incandescent lamp ends up as waste heat. CFLs use, on average, 20–25% of the energy to provide light at the same level and last at least 5 to 10 times longer, but the higher initial cost is the barrier to their wider use. However, a long-term view which considers replacement requirements of the two products suggests that the CFL is cheaper (ibid.). The cheapest way to conserve energy is to switch off lights when rooms are unoccupied.

Refrigerators and freezers

The amount of energy consumed by refrigeration appliances depends not only on the technical design of the appliance, but on the number of times the doors are opened, the amount of food loaded and the temperature of the items. The HEEP study recorded 30% less energy consumption than the manufacturers' label values. Therefore, though the label value could be used to compare appliances, it should not be used to estimate absolute energy consumption. Further, the number of occupants, age of the refrigerator, temperature of the surrounding area and whether the refrigerator is stand-alone or built-in were identified as factors influencing energy use (Stoecklein et al. 1998b). Compared with fridge–freezers, freezers use less energy (ibid.).

Domestic appliances

With the exception of refrigerators and freezers, the performance of all appliances depends on the user and the duration of use. According to HEEP, appliance energy use accounts for 24% of total energy use in the NZ house (Isaacs 2004: p. 7). Between 1990 and 2000 the energy used by appliances in the UK house increased by 9% (DTI 2002: p. 26) partly as a result of the increasing number of appliances used. Owing to the presence of electronic and computer controllers in appliances, standby power use, which can vary from 0 to 20 W depending on the appliance, is a growing concern. HEEP has recorded a 300 W continuous load attributable to fridge–freezers, clocks and to the standby load of appliances such as TV and stereo (Isaacs 1997). Although individually these may seem trivial, a combination of continuous loads could constitute a significant fraction of total energy consumption. In the UK, standby power is estimated to be responsible for 1% of total domestic energy use (DTI 2002: p. 27)

HEEP has demonstrated that 71% of the electrical energy consumption of an NZ house depends on the number of occupants, whether or not a hot water cylinder is insulated, the type and number of lights used and the level of household income (Stoecklein et al. 1998).

Life cycle energy

The embodied energy and operating energy issues of a house have been examined and established in the preceding sections. However, in order to establish the total energy attributable to a house over its lifetime both embodied and operating energy should be considered in concert. Life cycle energy is both embodied and operating energy attributable to the building throughout the lifetime of the building.

Recurrent embodied energy

Buildings have a longer life compared with most other products. Buildings start their life with a certain degree of fitness for purpose. However, the degree of fitness declines with age, wear and tear and the effects of weathering. The rate of this deterioration will depend on the durability of the materials used and the care of the users. Therefore, a durable material that lasts longer may provide a net energy saving even though it has a higher embodied energy.

During a useful lifetime, embodied energy is added to the building as it is maintained, refurbished or extended so that the building is able to provide the housing services that it was originally designed to do. Buildings may be maintained and upgraded to revised standards so long as this is cheaper than replacement construction. (Buildings may also be maintained for cultural/ historical reasons.) Timing of replacements depends on many technological, economic, legal, political and fashion related factors. The energy which is added to the initial construction embodied energy during the maintenance and refurbishment is known as the recurrent embodied energy. Reuse of building materials and elements will not only save much of this energy and cost of new construction, but eliminate the cost of demolition and disposal.

While rehabilitation and maintenance can reverse the depreciation of housing services they can also extend the life expectancy. However, expenditure on rehabilitation does not always guarantee an extension of useful life as buildings may sometimes be demolished shortly after extensive rehabilitation due to economic reasons.

The useful life of building materials and elements is decided based on two concepts: technical life and aesthetic life. Building materials and elements have to be replaced at the end of their technical lifespan as they wear out, but some materials and elements may be replaced before the end of their technical life owing to altered fashion or because the user wants a fresh appearance, as in the case of wallpaper and carpets. This represents the end of the aesthetic life. However, establishing the useful life of building materials and elements is an expensive and time-consuming task, which could sometimes lead to legal obligations owing to numerous environmental and use related factors. Therefore, building materials manufacturers generally attempt to satisfy the minimum regulatory requirements for their products (Haberecht and Bennett 1999).

Domestic appliances and equipment

Domestic appliances and equipment have a shorter useful life relative to the building structure and are replaced either when they no longer function properly, or due to market pressures to replace the existing with a newer model.

Furniture

The useful life of household furniture fitted and loose is estimated as 15.5 and 10 years respectively by Rawlinson (1998), while Fay (1999) estimated it to be 25 years. The embodied energy data for domestic appliances and furniture have not been included in the previous studies carried out in New Zealand and there is a need to establish figures representative of practices in the New Zealand context.

Demolition and recycling

The amount of energy required to demolish a house at the end of its useful life has been established as $10\,kWh/m^2$ by Adalberth (1997: p. 327). This is very small compared with the remainder of the life cycle energy. Energy gains from recycling and reuse depend on the collection and distribution system and the condition of the worn-out materials and on the energy required for cleaning for reuse or burning for fuel. However, the energy gain from recycling and reuse is attributable to the second use and not to the demolition of the building.

Conclusions

Energy attributable to buildings consists of: energy embodied in the materials and elements used in construction (embodied energy); energy to operate the buildings throughout their useful life (operating energy); and the energy added during the maintenance, renovation and replacement of the building materials and elements – the sum of all these is known as life cycle energy.

Owing to lower comfort expectations leading to lower internal temperatures and the relatively mild weather characteristics of New Zealand, space heating energy use is comparatively low. In the majority of NZ houses, more electricity is used for water heating than for space heating. The trend in space heating would be likely to be a rise in comfort expectations leading to a move from 'spot heating' to central heating over the years. There is an upward trend in the use of lighting associated with income. About one third of the total energy used in an NZ house is for operation of appliances and equipment, although the usage pattern varies widely. The operating energy and the maintenance energy of an NZ house are together several times greater than the embodied energy over its lifetime.

Based on the above information, a life cycle energy analysis was undertaken for a range of typical NZ houses – the details of this study are to be found in Part B. The next chapter examines the theory of life cycle costing, techniques of investment appraisal and application of life cycle costing to houses.

Notes

[1] Department of Trade and Industry, UK
[2] http://www.ecan.govt.nz/Our + Environment/Air/Air + Plan/ Proposed + New + Home + Heating + Rules/
[3] A gigajoule is roughly equivalent to the energy content in 30 l of petrol or 45 kg of coal.
[4] Centre for Housing Research, New Zealand
[5] http://xtabs.stats.govt.nz/eng/tablefinder/index.asp

References

Adalberth, K. (1997) Energy use during the life cycle of single unit dwellings: Example. *Building and Environment* 32(4): 321–329.

AS/NZS3580.9.10:2006 *Methods for sampling and analysis of ambient air: Determination of suspended particulate matter – PM2.5 low volume sampler – Gravimetric method.* Standards New Zealand.

Baird, G. and Chan, S. A. (1983 *Energy Cost of Houses and Light Construction Buildings* (Report No.76). New Zealand Energy Research and Development Committee, University of Auckland.

Barnett, D. L. and Browning, W. D. (eds) (1995) *A Primer on Sustainable Building.* Rocky Mountain Institute, Green Development Services.

Bassett, M. (1985) The Infiltration Component of Ventilation in New Zealand Houses, Ventilation Strategies and Measurement Techniques. *6th AIC Conference.* Netherlands, September.

Bell, M., Lowe, R. and Roberts, P. (1996) *Energy Efficiency in Housing.* Ashgate Publishing Company.

BS 5925:1991 *Code of Practice for Ventilation Principles and Designing for Natural Ventilation.* British Standards Institution.

Centre for Housing Research (2004) *Changes in the Structure of the New Zealand Housing Market, Executive Summary.* DTZ New Zealand.

Department of Trade and Industry (2002) *Energy Consumption in the United Kingdom,* Available at: http://www.dti.gov.uk/energy/inform/energy_consumption/ecuk.pdf [accessed 30 March 2006].

Department of Trade and Industry (2006) *Energy consumption tables.* Available at: http://www.dti.gov.uk/energy/inform/energy_consumption [accessed 30 March 2006].

Energy Information Administration. (2004) *Residential Energy Consumption Surveys: 2001 consumption and expenditure tables.* Available at: http://www.eia.doe.gov/emeu/recs/contents.html [accessed 31 March 2006].

Fay, M. R. (1999) *Comparative Life Cycle Energy Studies of Typical Australian Suburban Dwellings.* PhD Thesis, Deakin University, Australia.

Goodland, R. and Daly, H. (1996) Environmental sustainability: Universal and non-negotiable. *Ecological Applications* 6(4): 1002–1017.

Haberecht, P. W. and Bennett, A. F. (1999) Experience with Durability Assessment and Performance-based Building Codes. *First Asia Pacific Conference on Harmonisation of Durability Standards and Performance Tests for Components in Buildings and Infrastructure,* Thailand, September 8–10, 1999 (reprint *BRANZ Conference Paper No. 72, 1999*).

Harris, G., Gales, S., Allan, R. et al. (1993) *Promoting the Market for Energy Efficiency.* Energy and Resources Division of Ministry of Commerce, Institute of Economic Research.

Hendtlass, C. A. (1981) *Report on a Survey into Hot Water Usage Patterns in Residential Dwellings.* Joint Centre for Environmental Sciences, University of Canterbury, New Zealand.

International Energy Agency (2004) *Electricity Information.* IEA Statistics.

International Energy Agency (2005) *Key world energy statistics.* Available at: http://www.oecd.org/topicstatsportal/0,2647, en_2825_495616_1_1_1_1_1,00.html [accessed 27 March 2006].

Isaacs, N. P. (compiled & edited) (1997). *Energy Use in New Zealand Households* (Report on the Household Energy End Use Project (HEEP) – Year 1). Energy Efficiency and Conservation Authority.

Isaacs, N. P. (1998) Poverty and Comfort? *Fourth National Food Bank Conference,* Wellington, November 13, 1998 (reprint *BRANZ Conference Paper No. 59, 1998*).

Isaacs, N. P. (1999) Building Science, Regulations and Health. *EECA Seminar Improving Health and Energy Efficiency through Housing,* Wellington, Christchurch and Auckland, 13–15 September 1999b (reprint *BRANZ Conference Paper No. 73, 1999*).

Isaacs, N. (2004) Supply Requires Demand – Where Does All of New Zealand's Energy Go? *Royal Society of New Zealand Conference,* Christchurch, 18 November 2004 (reprint *BRANZ Conference Paper No. 110, 2004*).

Isaacs, N., Camilleri, M. and Pollard, A. (2004) Household Energy Use in a Temperate Climate. *American Council for Energy Efficient Economy 2004 Summer Study on Energy Efficiency in Buildings*, California, 23–28 August 2004 (reprint *BRANZ Conference Paper No. 108, 2004*).

Isaacs, N., Camilleri, M., French, L. et al. (2005) *Energy Use in New Zealand Households* (HEEP Year 9 Report). Building Research Association of New Zealand (*BRANZ Study Report No. SR141, 2005*).

Ministry of Economic Development (2001) *New Zealand Energy Data File – January 2001* (compiled by Hien, D. T. Dang). Energy Modelling and Statistics Unit, Ministry of Economic Development.

Ministry of Economic Development (2002) *New Zealand Energy Data File – July 2002* (compiled by Hien, D. T. Dang). Energy Modelling and Statistics Unit, Ministry of Economic Development.

Ministry of Economic Development (2005) *New Zealand Energy Data File – July 2005* (compiled by Hien, D. T. Dang). Energy Modelling and Statistics Unit, Ministry of Economic Development.

NZS4218:1996 *Energy Efficiency: Housing and Small Building Envelope*. Standards New Zealand.

Parker, G. J. and Tucker, A. S. (1985) *Dynamic Simulation of a Domestic Hot Water System*. Unpublished draft, Department of Mechanical Engineering, University of Canterbury, New Zealand.

Pilatowicz, G. (1995) *Eco-Interiors: A Guide to Environmentally Conscious Interior Design*. John Wiley and Sons Inc.

Pollard, A., Stoecklein, A. and Bishop, S. (1998) Preliminary Findings on the Internal Temperatures in a selection of Wanganui Houses. *Mites, Asthma and Domestic Design III Conference*, Wellington, 14–15 November (reprint *BRANZ Conference Paper No. 53, 1998*).

Rawlinson & Co. (ed.) (1998) *Rawlinsons: New Zealand Construction Handbook*. Rawlinsons New Zealand Constructions Handbook Ltd.

Statistics New Zealand (2002) *2001 Census of Population and Dwellings: Housing*. Statistics New Zealand.

Stoecklein, A., Pollard, A., Isaacs, N. et al. (1998) Energy End-Use and Socio/Demographic Occupant Characteristics of New Zealand Households. *IPENZ Conference*, Auckland, New Zealand, 1998 (reprint *BRANZ Conference Paper No. 52, 1998*).

Stoecklein, A., Pollard, A. and Bishop, S. (1998b) Energy End-Use in New Zealand Houses. *ACEEE Summer Study*, CA, USA (reprint *BRANZ Conference Paper No. 57, 1998*).

Trethowen, H. A. (1979) *The Need for Energy Research in Buildings* (Research Report R31). Building Research Association of New Zealand.

Trethowen, H. A. and Bassett, M. (1979) Prospects for Domestic Energy Conservation in New Zealand. *General Session Papers Forum Papers 4th New Zealand Energy Conference*, Energy and New Zealand Society, Auckland, 17–19 May 1979, pp. 124–129.

Vale, B. and Vale, R. (1991) *Towards a Green Architecture*. RIBA Publications Ltd.

Wright, J. and Baines, J. (1986) *Supply Curves of Conserved Energy: The Potential for Conservation in New Zealand's Houses*. Centre for Resource Management, University of Canterbury, New Zealand.

4 Life cycle costing of buildings

Buildings are judged by the value provided for the money spent, although they are also judged on their appearance and the way they function. However, cost has always been the first issue for many designers as the cost may well limit what can be achieved aesthetically, and can even affect how well the building functions. The initial cost of buildings may be reduced by limiting the built area, by adopting simple structural systems and suitable construction methods, and by designing for the use of standardised components. However, buildings have a long life and need to continue to provide their services over long periods, during which the building costs continue in the form of operating and maintenance costs, costs of adapting to changing needs over time, cost of inconvenience caused by the building for the function carried out within it, etc. Stone (1980) argued that the subsequent costs associated with buildings are about three times the initial cost of construction, while Flanagan et al. (1989) estimated these to be about 55% of the life cycle cost over a useful life of 40 years. These estimates vary widely, which may be because the actual relationship between initial cost and subsequent costs depends on the quality and performance expected of the building. Nevertheless, a proper balance between the initial construction cost and the subsequent operating and maintenance costs could lead to improved efficiency in the use of resources. A slightly higher initial cost might not only reduce the frequency of and the expenditure on maintenance, but it could result in an improved aesthetic quality and less disruption during the useful life of the building.

Investigation of building design team interactions in the UK has revealed that the client actions during the design stage are driven by the concern for initial cost, increasingly towards the latter stages of design (Wallace 1987). Although this situation might be expected to change over time, a more recent survey of the New Zealand construction industry in 1995 (Donn et al. 1995) has revealed that the client requirement is believed by New Zealand architects and designers to be for a minimum initial cost rather than a minimum life cycle cost (in terms of either finance or energy). Bird (1987) suggests that the interest of owners or occupants in the life cycle cost of the building would depend on the type (residential, retail establishments, offices, etc.), its use, market situation and overall circumstances of the building. However, it is also reasonable to expect this interest to vary depending on whether the building is for personal use, for rent or for sale. Haberecht and Bennett (1999) have argued that the ownership of New Zealand houses changes approximately every 7 years, in which case the life cycle cost becomes rather meaningless to the individual house owner.

However, as a society, New Zealand and other countries have recognised an international responsibility to improve resource efficiency according to the Kyoto Protocol requirements by stabilising greenhouse emissions at agreed levels (as discussed in the two sections in Chapter 1, Environmental effects of energy use and Climate change). Research (MED 2003: p. ix) has estimated that, at a GDP growth rate of 2.5% and with a carbon tax of 15NZ\$/t CO_2, from 2008 the carbon dioxide emissions from the energy and industrial sectors in New Zealand are expected to be about 16% above the level allowed by the Kyoto Protocol during the target period. Hence, interest in the performance of buildings and in the costs to be incurred over their useful life as a means of evaluating the alternative design/construction options, could be expected to rise over time. In any case, the consideration of life cycle is crucial for increasing the uptake of more sustainable building design practices that will continue to provide significant operating and maintenance energy and cost benefits over long periods. Life cycle costing (LCC) provides a means of comparing the initial and life cycle costs for such buildings.

Life cycle costing is an economic assessment of competing design alternatives in which all significant costs of ownership are considered over the useful life of each alternative and all are expressed in terms of equivalent dollars (Kirk and Dellísola 1995: p. 9). Therefore, this method provides an estimate of not only the initial construction cost, but the cost of maintenance and operation. Both maintenance and operating requirements as identified in Chapter 3 are required if a house is to satisfy the purpose for which it is intended and if the potential useful life is to be realised. The expenditure profile and the relationship between the initial construction cost and the life cycle operating costs vary between different construction types. Therefore, the object of using LCC is to evaluate the different options in similar objective terms. In practice however, this evaluation involves both quantitative and qualitative aspects, such as comfort, aesthetics and convenience, many of which are subjective. While the quantitative aspects provide the baseline reference for the assessment, many important factors that govern the final assessment may be qualitative, and professional value judgement has to be used to refine the final assessment. However, conventionally the qualitative assessment of design options is done without any quantitative data to provide alternative assessment criteria. Life cycle costing, therefore, adds a quantitative criterion to a situation where qualitative judgement is already being used.

During the 1930s LCC was mainly used for engineering economics. Life cycle costing appears to have been used in the acquisition processes for the US government from as early as the 1930s (Kirk and Dellísola 1995: p. 6). Since the 1960s the principles of LCC have been applied to buildings and in the USA all buildings have been evaluated based on life cycle costs since 1978 as a requirement of the National Energy Conservation Policy (Kirk and Dellísola 1995: p. 8). The principles of LCC have also been adopted in other cost evaluation techniques such as costs-in-use (Stone 1980), ultimate costs and terotechnology – a combination of engineering, financial, management and other practices applied to consideration of physical assets in order to achieve economic costs over their commercial life time (acquisition to disposal) (Seeley 1996).

The uses of LCC are numerous but lie principally in two main categories: evaluation of alternative options; and financial planning. The role of LCC in the building and construction industry can be defined as (among other renderings):

- an evaluation technique that can be used to choose between competing options based on total life cycle cost rather than the initial cost;
- a method to estimate and budget for future operating and maintenance costs;
- a system to assess new materials and technology;
- a tool to determine cost drivers; and
- a practice useful in reducing the total project cost.

Life cycle costing may be applied to the whole building, or to a specific element or a detail. When applied to whole buildings, LCC can highlight interrelationships between decisions and resulting cost trade-offs. While the life cycle cost of a complete building is derived by summating the life cycle costs of individual building elements, LCC may also be used to select individual building components/elements based on the overall life cycle cost. Care must be taken where LCC is used for individual elements, as elements may appear less cost-effective when taken out of their context as part of the whole building. Life cycle costing is more valuable during the earlier phases of a project such as concept development, preliminary design and design development, owing to the greater potential for savings. During the early stages changes can be made with minimum effort and optimal effect. Once the construction is started the cost of change increases rapidly. The bulk of the expenses during the operation phase of the project depend on the decisions taken in the early phases of the design in terms of the need for the project, operating requirements, support concepts, etc. The results of LCC carried out during construction and occupancy stages would, however, provide feedback data for future projects of a similar nature.

The basic methodology used in the comparative analysis of life cycle costs of buildings consists of:

- identification of the elements that are common to the options being considered and removal of such items from the comparison;
- identification of significant costs associated with each of the options being considered;
- addition of each group cost by the year in which it is incurred and discounting to a common base or to present worth;
- selection of the lowest cost option;
- checking the effect of the assumptions used on the final result using a sensitivity analysis; and
- tempering the final selection with non-economic considerations such as aesthetics, safety and the environmental considerations (Kirk and Dellísola 1995: p. 11).

In LCC, the design alternatives are compared in relative rather than absolute terms, as the prediction errors are usually relative rather than absolute (Stone 1980: p. 40). While this relative comparison highlights the apparent relative cost differences between the design alternatives, which would otherwise be concealed in an absolute comparison, the removal of common elements from the analysis reduces the time and the complexity of the analysis. However, it is important that the common elements are not assumed to be unchangeable if in fact this is not the case. When LCC is used for financial planning the complete range of cost factors should be analysed.

Depending on the nature of the project and the objectives of the client, the period of analysis used in LCC can be either the entire useful life of the building or a definite period over which the client is interested in the building. In the USA, federal buildings are analysed for a period of 25 years, which is much shorter than the useful life of those buildings (Flanagan et al. 1989: p. 35).

In life cycle cost analysis, present time marks the beginning of the period of analysis for which the base costs are quoted. The common choices for the present time include, the design stage, halfway through the construction phase and the beginning of occupancy. In economic analysis systems commonly used for LCC of buildings, the beginning of occupancy is used as the present time (Kirk and Dellísola 1995: p. 30).

Like life cycle analysis (LCA), LCC also depends on operating costs and performance data based on past experience and forecasts of future events. Forecasts are based on the assumption that future trends can be predicted, to some extent, based on past patterns, which is true only when there are no significant changes in the trend – such as changes due to new discoveries or a technological leap. The prediction of future annual inflation rates, and taxation or energy price increases, is surrounded with uncertainty. The increase in oil prices globally during 2006 would have been difficult to forecast using any forecasting methodology. Therefore, LCC is not an attempt to predict accurately the actual cost, as the many assumptions used to derive the life cycle cost can differ from the actual situation during the subsequent stages of building life. Although data quality and methodological imperfections are common shortcomings in both LCA and LCC, their use should still improve the basis upon which decisions are made.

Lack of cost data and industry standards on life cycle behaviour of buildings, short-term focus on cash flows and the diverse focus of numerous members involved in the design and construction of buildings have been identified as barriers to wider implementation of LCC in the construction industry (Abraham and Dickinson 1998). A survey of Swedish building developers and clients in 1999 found that the lack of experience in using LCC models and the complexity of available models together with the lack of data on new materials and operating systems further constrained the use of LCC in the construction industry (Sterner 2000). Although operating and maintenance requirements are context specific, Ferry and Flanagan (1991) argued that the expert judgement on maintenance and operating requirements of buildings would be sufficient for LCC when used for selecting the best alternative, as LCC is only a means to an end. However, this would not be practical in the case of new materials and innovative construction methods and technologies, such as those that might be used in more sustainable designs, owing to the deficiency in information and experience. Further, life expectancy of many building components and fittings may be far shorter or longer than what is assumed in these calculations. This applies to 'technological' components such as chillers and glazing units, but may apply also to things that do not wear out but that become unfashionable, and which therefore are thrown out in a style-based refit. Gluch and Baumann (2004) have highlighted the inability of LCC to handle environmental implications as a result of the exclusion of common goods and cost to future generations in LCC calculations, together with the conflicting nature of irreversible decisions, which is contradictory to common economic theory. Economic theory assumes that the decision-making is rational and all consequences are known prior to selection from a sample of competing options. This, however, is contradictory to the uncertain, complex and long-term nature of environmental decisions.

Certain economic concepts such as the time value of money, discounted cash flows and discount rates, are essential for LCC. These are discussed next.

Time value of money

A sum of money spent or received at various points in time has a different value. The value could be larger as a result of interest payments or smaller because of effects of inflation at a future date. This ability of money to earn money, and thus to increase over time, is known as the time value of money (Kirk and Dellísola 1995: p. 19). The life cycle cost consists of a number of sums of money invested at various points in time. Therefore, to calculate the life cycle cost of a house it is incorrect to add up the face value of these expenses as and when they occur. In order to convert expenses occurring at various points in time to a common basis, the concept of present value (also known as present worth) is used. As a result of the effects of interest earned, the present value of the cash flows is less than their forecast or future value. The concept of present value is similar to the concept of compound interst applied in reverse. This is further explained, as follows.

With an interest rate of $r\%$ an expenditure of P today is equivalent to an expenditure of $P(1+r)$ one year in the future. (Note: r is expressed as a fraction of 100; for example 6% becomes 0.06.) If the expenditure next year (commonly known as terminal expenditure) is T, then

$$T = P(1 + r)$$

Therefore, the present value P, of the terminal expenditure T is given by

$$P = \frac{T}{(1 + r)}$$

The present value of a future sum is the sum of money that should be set aside today to cover expenditure in the future (Flanagan et al. 1989: p. 24). Therefore, it is the present exchange value for a future sum at the given interest rate. Hence, the present value P_n of a terminal expenditure T_n in n number of years at an interest rate of r is given by

$$P_n = \frac{T_n}{(1 + r)^n}$$

Similarly, the present value of a series of annual payments is given by

$$P_n = A \left[\frac{1 - (1 + r)^{-n}}{r} \right]$$

where A is the annual payment.

This method of calculation is known as discounting, as future costs are discounted to a lesser value when converted to the present time. The present value of a future sum is lower the further away in time the sum is due. Therefore, expenditure occurring beyond about 25 years is unlikely to make any significant difference to the ranking of options (Flanagan et al. 1989: p. 40) in a life cycle cost analysis. The use of discounted cash flows over the life cycle for evaluation of commodities such as cars and computers is a well-established economic practice. Buildings have a very long lifetime compared to these commodities and when applied to buildings the use of discounted costs tend to be biased towards the initial cost and to diminish the significance of maintenance and operation costs. However, most sustainable design and construction practices tend to provide benefits in terms of lower maintenance and operating requirements which are not emphasised by discounted costs. This problem has already been stressed by researchers such as Awerbuch (1993), De Brito and Franco (1994) and Nicolini et al. (2000).

Inflation

Inflation is the general increase in price of the same goods and services over time, i.e. an increase in cost without an increase in value (Kirk and Dellísola 1995: p. 25). The cause of inflation is too much money chasing too few goods. Inflation is of two kinds: general inflation and differential rates of inflation. General inflation affects the time value of money. Differential rates of inflation apply to items that may inflate at rates higher than the general inflation, such as energy prices.

Costs in LCC

In LCC, costs are the raw material of analysis and the basis on which the competing options are evaluated. However, cost is only a single dimension in terms of a particular situation or a building. Costs associated with a building reflect the physical nature of the building, performance expectations and the quality maintained within the building. Costs may also be affected by the location of the building in terms of taxes and fees, and, also, sensitivity to environmental impacts. Therefore, the selection of historical cost data to be used in a particular situation should take all the above into account.

Costs are of two types: tangible costs and intangible costs. Tangible costs are those which are quantifiable in monetary terms, such as initial costs and operating costs. Even though it is difficult to quantify intangible costs in monetary terms, in some cases these costs, such as desirability created by aesthetics and cost of denial of use owing to the need for maintenance work, may play an important role in the final decision.

Although the initial cost is the largest single cost for most building types, it has been estimated to be less than 50% of the total life cycle cost (Flanagan et al. 1989: p. 9). Stone (1980) determined the composition of the life cycle cost of a conventional house at a discount rate of 5% to be as follows.

- Initial construction cost   56%
- Maintenance cost   16%
- Fuel for heating and lighting   28%

Almost half of the above initial costs arise from fittings, finishings, equipment and site works, which are necessary irrespective of whether the building is permanent or temporary. These figures suggest that a long life for houses is essential if life cycle costs of housing are to be reduced. The order of the figures may be altered by changes in design objectives; for example, initial construction cost might be increased to secure lower maintenance and fuel costs. Although furniture and appliances are not included in the above list, a life cycle study of New Zealand houses that took them into account found that that they are significant relative to the total cost because of their relatively short useful life (Mithraratne and Vale 2004).

The composition of recurrent costs, also known as costs in use, varies among different types of construction and building. The general ranges of these costs as a percentage of the total, according to the Joint Centre for Land Development Studies (n.d.: p. 2/1) for commercial buildings, are as follows.

- maintenance   7–30%
- energy   15–45%
- cleaning   5–40%
- rates   5–45%
- insurance   2–20%
- security and management   0–10%

Major maintenance costs are incurred when building elements/components are replaced as a result of failure. The future cost of such work is expected to increase relative to the prices for new construction work and to further increase proportionately, the greater the element of labour relative to material. In any case, maintenance work tends to be expensive compared with initial construction work owing to a reduction in the scale of work, the need to strip out the old work, the need to work in confined spaces and the need often to work overtime when the building is being extensively used. Stone (1980: p. 61) has estimated that, for most building types, maintenance costs per floor area average 1.5% of the initial costs at constant prices. While maintenance costs occur intermittently in the useful life of a building, all the other costs identified above, being annual costs, are evenly distributed.

As identified by Kirk and Dellísola (1995: p. 9), life cycle costs of building and development activities include the following elements.

- initial costs – project costs, construction costs and other costs;
- financing costs – any debt associated with the initial capital cost;
- operating costs – cost of energy, water and other 'utilities';
- maintenance costs – cost of regular custodial care and repair;
- alteration and replacement costs – costs relating to changing the function of the space (alteration cost) and costs to maintain original function (replacement cost);
- taxes, credits and depreciation based on current tax laws;
- associated costs – cost of staff, materials, etc., for the function that is to be performed, denial of use due to major maintenance and refurbishment, security and insurance;
- salvage value – positive if there is a residual value and negative if demolition is required.

In terms of initial costs, items such as cost of land acquisition, cost of demolition of existing structures and cost of site preparation also need to be considered in life cycle cost evaluations. Cost of equipment/appliances, furniture and furnishings, etc. that need to be replaced at relatively short intervals could add significantly to the operating costs. Although the cost of demolition and disposal of demolition debris is not included in the above list, depending on the location of the disposal sites and the quantity of materials involved, this may become a major cost. The inclusion of demolition and disposal costs could be vital, particularly for materials associated with health and environmental concerns, such as asbestos cement sheeting, because of the higher disposal costs involved. Materials that are difficult to recycle could also lead to higher than normal disposal costs. Adalberth (1997: p. 327) has estimated the transport energy requirement for demolition waste to be 20–$30\,kWh/m^2$ of floor area for residential buildings based on a distance of 20 km from site to landfill.

The methods used for estimation of costs relevant to the LCC analysis depend on the availability of information. Factors such as the newness of the material or technology used, the degree of certainty of the use thereof and the useful life expectation, affect the reliability of the information available. Also, the availability of the information depends on the stage of the life cycle at which the analysis is undertaken. The following are the basic methods used for estimating costs in LCC, depending on the descending order of availability of information (Standards New Zealand n.d.: p. 14).

- Engineering cost method – direct estimation of cost by examination using standard established cost factors. Provides an accurate estimate and is used when most of the information is known.
- Analogous cost method – estimation based on experience. Uses historical data updated to reflect real costs to the base date or the advancement of the new products. Accuracy depends on the relevancy and appropriateness of the analogy used.
- Parametric cost method – use of parameters and variables to develop cost–element relationships in the form of equations. This reflects the analyst's assessment of the way the costs are generated.

Costs can also be categorised as internal costs and external (social) costs (Stone 1980: p. 10). Internal costs are those costs borne by the building user while the external costs are the costs borne by other individuals and the community at large, as a result of the design of the building or the way it is used. While the price paid for the building site represents the real cost to the client, to the community the cost would be the value of the output lost as a result of using the site in one way rather than in another. In trying to produce the design most satisfactory to the requirements of the individual client, cost may be generated which other members of the community have to bear – such as traffic congestion and accidents, air pollution, noise, and even unsightliness. However, because some individual clients may not wish to increase their costs to ensure better use of national resources, regulatory measures such as by-laws, taxes and subsidies generally have to be imposed to protect the community. Life cycle costing evaluations usually ignore these external costs borne by the wider society. However, buildings use large amounts of resources (such as land, energy, and materials,) and continue to generate emissions over long periods. Therefore, the social costs associated with construction, use and disposal of buildings could be significant.

Although LCC deals with the costs, for proper evaluation of options being considered both costs and benefits of each option should be considered in the LCC exercise. As with costs, benefits may also be tangible and intangible. Therefore, all the benefits of an option need to be listed and quantified in monetary terms and included in the LCC as an appropriately discounted negative cost. When the benefits – such as comfort, convenience and prestige – cannot be expressed in monetary terms, professional value judgement is used to rank the options with respect to the relative benefits, as is done for intangible costs.

Real costs, nominal costs and discounted costs

The basic concept used in LCC is the time value of money. As discussed earlier, owing to the effects of inflation and the general growth in costs, the cost of an item varies depending on when the cost is incurred. The costs in an LCC are expressed in different ways, depending on the purpose of the analysis, as real costs, nominal costs and discounted costs.

Real costs

Real costs are the costs measured in terms of resources or in terms of each other (Stone 1980: p. 52). The real costs are not affected by any inflation, as the same but no more materials, labour, and time are needed to complete an operation before as after a period of inflation. However, the value represented by the real cost would be the amount that would be due if the costs were incurred at the base date and not at a future point in time. Since the face value of costs (or the purchasing power of money) varies depending on when costs are incurred as a result of inflation, and the change in costs would not be the same for all the items, the real costs are always tied to a base date. Although real costs are independent of inflation, differential price escalation caused by technological advancement and efficiency improvements, which will probably occur at a rate different from general price escalation, needs to be incorporated in real costs and the costs adjusted accordingly.

Real costs provide accurate comparisons as current values are used, and the need to predict inflation is eliminated. However, they are inappropriate for financial budgeting, when actual amounts of money today are required to ensure that the actual amount needed for the future expenditure is secured. The basic problem is that the future costs expressed in real costs do not provide a true picture.

Real costs may be used:

- to identify the major elements of cost and cost-effective improvements;
- to understand product design;
- to increase the user awareness of the costs and benefits of products, and for products with shorter lives; and
- to manage costs of a product through its design, construction and operation (Standards New Zealand n.d.: p. 8).

Nominal costs

Nominal costs represent the actual cost over the useful life expressed in terms of actual dollar amounts to be paid at a specific time. These costs are derived based on projected economic, technological and replacement factors. Nominal cost is the real cost subjected to expected general price inflation and the cost growth between the base date and the time when the costs occur. When future costs are predicted in terms of nominal costs, this represents the money that is needed at a future date to execute the maintenance. Although real costs may be added, as they are all related to a base date, nominal costs that occur across a period of time should not be added.

Nominal costs may be used:

- to allocate future funds for the product being developed;
- for long-term financial planning and budgeting of capital and operating costs; and
- to record data on price performance (Standards New Zealand n.d.: p. 8).

Real and nominal costs are related to each other by the general price inflation that occurs in between the dates considered. Therefore, the real costs could be converted to nominal costs by multiplying by an inflation factor of

$$f = (1 + a)^y$$

where a = expected increase in general prices per annum, and y = number of years between the base date and the date the cost occurs (Standards New Zealand n.d.: p. 23).

Discounted costs

Discounted costs are the future costs discounted using an appropriate discount rate, in order to convert expenses occurring at various points in time to a common basis. Depending on whether the costs to be discounted are expressed as real costs or nominal costs, the discount rate varies. While inflation is included in the discount rate when the nominal costs are used, inflation is excluded from the discount rate when the real costs are used in LCC.

Discounted costs may be used:

- to evaluate products with long lives using comparative analysis of alternative design approaches;
- to assess the impact of new technology;
- to evaluate and compare alternative strategies for product use, operation, inspection, maintenance, etc.;
- to evaluate and compare different approaches for replacement and rehabilitation of ageing facilities;
- to select from among competing tenders for products with long lives; and
- to enable a common basis for comparison for products with significantly different cost profiles (Standards New Zealand n.d.: p. 8).

Real costs that occur at a future time could be converted to discounted costs by multiplying by a factor of

$$f = \frac{1}{(1 + d_r)^y}$$

where $d_r =$ the real rate of discount per annum, and $y =$ number of years between the base date and the date the cost occurs,

or

$$f = \frac{(1 + a)^y}{(1 + d_n)^y}$$

where, $a =$ the expected increase in price per annum, $y =$ number of years and $d_n =$ nominal discount rate per annum (Standards New Zealand n.d.: p. 24).

Similarly, nominal costs could be converted to discounted costs by multiplying by a factor of

$$f = \frac{1}{(1 + d_n)^y}$$

where, $d_n =$ the nominal discount rate per annum, and $y =$ number of years,

or

$$f = \frac{1}{(1 + a)^y (1 + d_r)^y}$$

here, $a =$ the expected increase in price per annum, $y =$ number of years, and $d_r =$ real discount rate per annum (Standards New Zealand n.d.: p. 24).

Current dollars and constant dollars

The price of an item at any time is the amount of dollars required to purchase the item at that time. This is the price in current dollars. When the price increases, the cost in current dollars also increases. Therefore, when a future cost is to be predicted in current dollars, the effects of inflation and cost growth should be included. If forecasted accurately this future cost would represent the actual amount of dollars needed in the future, in current dollars at that time.

However, as a result of the difficulties involved in accurately predicting the annual inflation rates and cost growth, future costs may be stated in terms of dollars of constant power in economic analyses. Constant dollars refer to the purchasing power a dollar had in a particular year (Kirk and Dellísola 1995: p. 28). Although the use of constant dollars to represent future costs is not realistic, the effects of inflation can be ignored in the analysis. Since LCC is mainly used to compare options, the constant dollar approach is commonly used.

In this approach, the present prices are only modified for the items that escalate in price over and above general inflation.

Discount rate

Discounting highlights the importance of the timing of cost and benefit flows. Therefore, the discount rate represents the time value of money, which depends on inflation, cost of capital, alternative investment opportunities and personal consumption preferences. Discount rate is the rate at which future money declines in value. Therefore, discount rate represents the value attached to costs and savings in the future. However, the final results of the LCC exercise are greatly dependent on the choice of the discount rate. Discount rate would depend on the type of investor. A private investor may accept the bank rate as the return for investment while a business would expect a relatively higher return, whereas the rate of return for public investment usually tends to be lower.

As future costs are discounted to a lesser value when converted to their present value, the interest rate is known as the discount rate (Flanagan et al. 1989: p. 9). The discount rate used could be either the nominal rate of increase in the value of money over time or the actual rate of increase in the value of money – i.e. the rate over and above the general inflation rate in the economy (Kirk and Dellísola 1995: p. 27).

The Joint Centre for Land Development Studies (n.d.: p. 2/27) have argued that the inflation of future costs would be matched by the inflation of money that would be available when such future costs are to be incurred and therefore the effects of general inflation may be omitted from LCC analysis. However, Flanagan et al. (1989: p. 28) have argued that the effects of inflation should be included in the discount rate as inflation has become an important factor in determining future costs. The discount rate used would depend on whether nominal or real costs were being discounted. With nominal costs the discount rate used (known as the nominal discount rate) should include a component for inflation while with real costs the discount rate used (known as the real discount rate) should not include a component for inflation.

The selection of the relevant discount rate may depend on the way the project is funded and the client's expectations:

• Cost of borrowing money – the discount rate depends on whether the project is financed through borrowing or from capital assets. If borrowed money is used, the discount rate should represent the cost of borrowing. This would be an indication of the market value of money over time.

- Minimum attractive rate of return – this discount rate includes an increment for the risk associated with the project in addition to the cost of borrowing.
- Opportunity rate of return – when capital assets are used to finance the project the discount rate depends on the best alternative use of the funds. This discount rate is the actual earning power of money.
- After inflation rate – private investors expect a certain rate over and above the general rate of inflation for the investment. This discount rate is equivalent to the average rate of return in the private sector less the inflation rate (Kirk and Dellísola 1995: p. 28). This discount rate is used when the costs are expressed as real costs (as effects of inflation are omitted from the calculation).

It is preferable to receive income as early as possible, and to defer expenditure as long as possible. Discount rate is the interest rate that would make it worth while to spend money in a year's time (thereby accruing a year's interest) rather than now. While a higher discount rate favours low capital cost alternatives with higher maintenance/operating costs, a low discount rate will favour future cost savings. Therefore, when a high discount rate is used, the future costs will have a lesser impact on the selection.

The use of a zero discount rate indicates that the timing of the expense/income is not significant. Also, the use of a zero discount rate is suggested as a method to account for definite negative environmental impacts by Gray et al. (1993). If not discounted, items such as non-renewable energy use would make a significant contribution to the total life cycle cost.

According to Standards New Zealand (n.d.: p. 24), the Australian government uses a real discount rate of between 5 and 9%, although many international studies use 3 or 4%. While public authorities are normally able to borrow money at the pure rate of interest plus an allowance for inflation, private clients have to pay a risk premium on top of this. While this risk premium is small for buildings which satisfy the requirements of most users and are therefore likely to be easily marketable, such as houses, offices, shops and normal factory buildings, Stone (1980: p. 59) has estimated this risk premium is likely to add about 2–3% to the rate of borrowing.

An interest rate consists of the real earning power of money and the effects of inflation. Therefore, an inflation-free discount rate may be calculated using the following equation.

$$d^l = \left[\frac{(1+d)}{(1+i)}\right] - 1$$

where, d^l = real discount rate (net of inflation discount rate), d = interest rate including inflation (nominal discount rate), and i = inflation rate (Flanagan et al. 1989: p. 28).

If the results of LCC are to be meaningful, the effects of both inflation and cost growth should be included, particularly if the effect on some options is greater than on the others. However, as future rates of inflation cannot be predicted with any accuracy over long periods of time, the inclusion of a realistic allowance for general inflation in building LCC is impossible. Energy and labour costs are likely to escalate faster than the capital cost and therefore differential inflation of such items cannot be ignored in LCC (Joint Centre for Land Development Studies n.d.: p. 1/1). Even though the differential rates of inflation cannot be predicted accurately, the acceptance of their presence could be used in the selection of alternatives. Although all the options are equally affected by inflation, its impact is greater on options for which the future cost to initial cost ratio is the largest.

Flanagan et al. (1989: p. viii) argued that avoiding surprises is a priority of most clients. Acceptance of the higher rates of certain items could be used at the design/construction phase to buffer the client from unexpected future expenditure. The effect of the change in the useful life and discount rate assumptions declines when the initial cost to running cost ratio is low rather than when it is high.

Therefore, the results of LCC depend on the approach, discount rate selected, how inflation is taken into account, the period of analysis, the beginning of the period of analysis and the types of costs included and/or ignored in the analysis.

Generally, design decisions related to residential constructions are driven mainly by factors such as appearance, functional requirements, personal preferences, fashion and prestige. However, in the case of decisions related to commercial buildings, more rigorous financial appraisals are warranted. These investment appraisal methods are discussed next.

Investment appraisal methods used for LCC

Investment appraisal methods facilitate the comparison of options being considered. Several methods are used for this purpose, some of which use non-discounting methods while some use discounting techniquess. Non-discounting methods should be used only over very short analysis periods, as future costs/savings could be disproportionately large compared with the initial cost.

The most commonly used investment appraisal method is the payback period technique (payback period is the time required to return the sum invested). Depending on whether the time value of money is considered or not, it is known as either the discounted payback method or the simple payback method, respectively. Although it is a quick and simple method easily understood by managers (as the speed of return rather than the rate of return is used), it is not a suitable technique for LCC, especially in the evaluation of more sustainable designs, as the costs beyond the period of payback are not considered. However, the method is useful as a means for initial screening.

Since LCC involves money spent over various points in time, interest formulae and investment appraisal methods are used to rank the options. Investment appraisal methods used for LCC should have the following qualities (Flanagan et al. 1989: p. 22).

- Cash-flows throughout the useful life of the building should be considered.
- Time value of money should be accounted for.
- Rate of return on the investment rather than the time should be considered (the rate should not be less than the market rate of interest).

The most commonly used investment appraisal methods, which take both the time value of money and the rate of return on the investment into account and which are used for LCC, are the net present value method and the annual equivalent value method (Kirk and Dellísola 1995: p. 24). Irrespective of the investment appraisal technique used, the costs for all of the options being compared must be calculated on the same basis; i.e. if the costs are expressed in nominal terms, this should extend to all of the options being considered so that the comparison is on a 'like-for-like' basis.

Net present value method

The net present value (NPV) of an item, system or a facility is the present value of the total investment to which one is committed as a consequence of a particular choice. Therefore, this represents the value of the sinking fund that has to be established today to cover all the costs that will be incurred throughout the useful life (Flanagan et al. 1989: p. 27). Hence, in this method of investment appraisal, the NPVs of future costs calculated over the useful life are discounted from the date on which they occur back to the present time and then added to produce the NPV of the life cycle cost of the project. The option with the lowest NPV would be the best in terms of life cycle cost. The main drawback of the NPV method is the difficulty in interpreting the results in a form that is meaningful to the clients.

When the replacement cycle of a component is not an exact multiple of the useful life of the building (or the period of analysis), a certain number of years of useful life of the component remain at the end of the period of analysis. Therefore, a positive residual value should be assigned in the NPV method to cover this remaining useful life of the component. An alternative means of handling this problem is afforded by use of the annual equivalent value method.

Annual equivalent value method

This method is also known as the annualised cost method and the equivalent uniform annual cost method. It combines the best attributes of the NPV method with the annual cost figures, producing results which are particularly useful to clients and building managers.

The annual equivalent value is the regular annual cost, when discounted, that just equals the NPV of the investment. Therefore, the annual equivalent value method calculates the NPV over the period of analysis and rather than expressing this figure as a one-time present value cost, it is then divided by the period of analysis to produce an equivalent uniform annual cost which includes both the effects of discounting and the larger items such as component replacement. The lower the annual equivalent value, the lower the total life cycle cost. This method is used with the assumption that there is a natural replacement cycle over which the materials/components are replaced by the same material, for example worn carpet being replaced by new carpet rather than by floor tiles, leading to identical consecutive costs.

The annual equivalent of the initial cost depends on the rate of interest to be paid on the initial cost borrowed, the useful life of the building and the initial cost itself. The higher the interest rate and the shorter the useful life, the greater the annual equivalent cost. The proportional effect of a change in the rate of interest on the annual equivalent cost is less for a building with a shorter life than for a building with a longer life. With the increase in the life of the building, the annual equivalent cost decreases until a relatively stable equal cost is reached. At the normal rates of interest paid for the money borrowed on buildings, the annual equivalent cost is about double for a short-life building (20 years) compared with what it would be for a building with a long life (over 60 years) (Stone 1980: p. 13). The lower the interest rate, the greater the difference in annual equivalent costs.

While NPV is useful to designers and developers who think in terms of capital costs, annual equivalent value is useful to property owners and building users who think in terms of annual costs.

Depreciation is an economic consequence of deterioration and obsolescence. Buildings do wear out and the capital is thereby eroded. According to Flanagan et al. (1989: p. 44), the rental value of an unrefurbished 20-year-old building is only about 60% of that of a new building and the capital value may be only about 50% of its modern equivalent. Although this could suggest that buildings have to be refurbished to suit the current fashions in order to guarantee market rent, lower life cycle cost could also be achieved from periodic maintenance without any refurbishment. This indicates that the current trend for architectural fashion to change rapidly can impact negatively on sustainable outcomes. Life cycle costing does not consider depreciation when calculating NPV or the annual equivalent amount. However, depreciation should be considered in the decision-making process.

Prediction errors and risk management

Since LCC involves the future, which is uncertain, and the data used are subjected to forecasting, estimation and assumptions based on current knowledge, the impact of changes on the results of LCC is likely to vary markedly. Although the way the buildings will be used or will function in the future is not known for certain, if the assumptions used for LCC are explicit, their effect of these assumptions on the final result can be analysed. The accuracy of the estimated life cycle cost also depends on the prediction errors, which are different from the mistakes in the calculations resulting from incompetence and carelessness. Prediction errors are measurement errors, sampling errors and assumption errors. While measurement errors arise from limitations in measuring techniques, the sampling errors arise by virtue of the fact that no sample is ever completely representative of the population from which it is drawn. Assumptions are always uncertain, and there may be many other assumptions possible that were not considered in the analysis. The largest source of error lies in predicting changes in future costs and the frequency with which operations need to be performed.

According to Flanagan et al. (1989: p. 125), a marked variation in the result of the life cycle cost analysis could occur as a result of a reasonable change in any of the following factors.

- the period of analysis
- the discount rate
- the life expectancy of the material/element
- the estimated cost
- the rate of inflation.

While an option with a high initial cost and low maintenance cost may become more attractive with a longer period of analysis, an option with a low initial cost may be preferred if a shorter period of analysis is used. Higher discount rates favour low initial costs and higher future costs, while lower discount rates have the reverse effect of increased relative preference for future cost-savings. Use of a low discount rate is similar to increasing the present value of future cost commitments. Selection of the option may also be affected by variations in the replacement cycles.

Sensitivity analysis

Sensitivity analysis, which forms a part of the general risk management system, is an important component of LCC. Sensitivity analysis identifies the extent to which a particular result is dependent on any change in key variables such as assumptions and estimates. As the results of LCC are used to influence the final choice from the options being compared, the sensitivity of the results of LCC – whether NPV or the annual equivalent cost – to the changes in any of the variables should be determined for proper analysis. The sensitivity analysis identifies the impact on the NPV of a change in a single parameter used in the LCC to calculate the NPV. However, the underlying assumption used in sensitivity analysis is that the nature of uncertainty is known. Sterner (2000) identified energy prices, discount rates and useful life as the dominant parameters used for sensistivity analysis in buildings LCC.

While there is no single method that is currently accepted as being the ideal for use with LCC, Flanagan et al. (1989) suggested the use of a spider diagram would allow the identification of the parameter most sensitive to a slight change (Figure 4.1). As detailed by Flanagan et al. (1989: p. 75), the methodology involved is as follows.

- Calculate the life cycle cost of a option using best estimates.
- Identify the parameters that are subject to risk and uncertainty.
- Select one of the identified parameters and recalculate the life cycle cost varying the parameter by $\pm x\%$ (x should be within the range of variation identified).
- Plot the results on a graph with *LCC* on the x axis and *percentage variation in the parameter* on the y axis.
- Repeat the exercise for the remaining parameters subject to uncertainty.

Each line in a spider diagram depicts the impact of a defined percentage variation of a single parameter on the life cycle cost. The flatter the line, the more sensitive the life cycle cost to the variation in that parameter. Therefore, this method could be used to identify the most sensitive parameter which affects the life cycle cost estimate and the ranking of competing options. However, this method does not provide any indication of the likely range of the variation in the parameter and also assumes that only one parameter is being varied at a time. In reality several risky parameters may vary simultaneously.

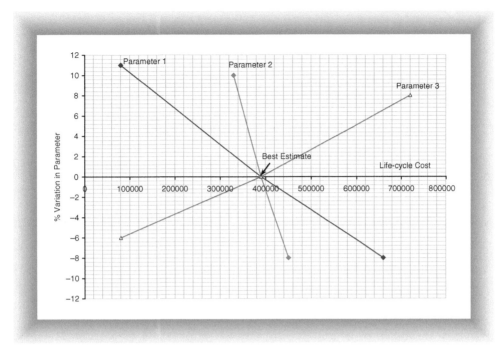

Fig. 4.1 Spider diagram for sensitivity analysis.
(Based on: Flanagan, R., Norman, G., Meadows, J. and Robinson, G. (1989) *Life Cycle Costing: Theory and Practice*. BSP Professional Books, p. 76.)

Flanagan et al. (1989) argued that the uncertainty of the range within which the parameter varies could be handled by defining the level of probability that a certain parameter would lie within a certain range. If the levels of probability are defined in this way for all the parameters, it would create probability contours as in Figure 4.2, which denote estimates of the likely range for variation of the life cycle cost. However, these contours are only subjective estimates and although a spider diagram provides information on factors which affect the final cost estimates, it does not provide any guidance on the selection of alternative options.

Although competing alternatives could be plotted on the same diagram with probability contours as in Figure 4.3, this would only provide an idea as to which parameter is more sensitive to change and the likely level at which one alternative may become more expensive in life cycle terms compared with the other. The greater the extent to which the ranking order is likely to change due to a variation in the parameter within the estimated probability contour, the less definite is the advice provided by the spider diagram for rejecting one alternative rather than another. In any case, sensitivity analysis assumes that the parameters are univariate, although, in reality, parameters are multivariate.

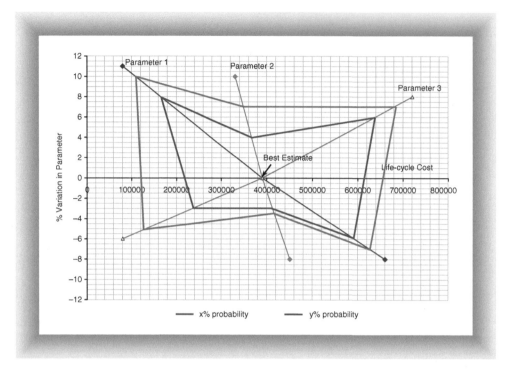

Fig. 4.2   Spider diagram with probability contours.
(Based on: Flanagan, R., Norman, G., Meadows, J. and Robinson, G. (1989) *Life Cycle Costing: Theory and Practice*. BSP Professional Books, p. 77.)

Therefore, while the spider diagram provides a quantitative method for determining the sensitivity of the life cycle cost to the variation of any parameter, the final selection of the option has to be based on value judgement. However, the above method provides a quantitative criterion on which to base the value judgement. Although probability analysis is a method which can be used to analyse change in several parameters simultaneously, it is a more complicated method which needs computer simulation, and although the method provides an overall assessment of the risks involved it is not possible to use it to evaluate the effect of change in an individual parameter.

Conclusions

The client requirement is usually perceived by designers to be for a minimum initial cost design rather than a minimum life cycle cost. However, owing to international commitments, such as Kyoto Protocol targets, a change in this situation could be expected in the near future.

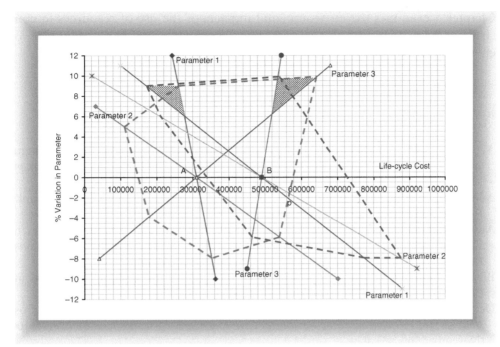

Fig. 4.3 Spider diagrams with probability contours for comparative analysis of alternative options.
(Based on: Flanagan, R., Norman, G., Meadows, J. and Robinson, G. (1989) *Life Cycle Costing: Theory and Practice*. BSP Professional Books, p. 78)

Life cycle costing provides a means of evaluating different construction and design options based on the total life cycle cost. The underlying concept used in LCC is the ability of money to increase in amount over time, known as the time value of money. Since future costs are lower in value when converted to their present value, a discount rate is used to convert all the expenses occurring at various points in time to present value. The discount rate depends on the objectives of the client and the way the project is financed.

Costs form the basis of evaluation in LCC. However, both costs and benefits associated with the options being compared, expressed in equal terms, and with or without inflation, should be considered for proper evaluation of the options. Inflation is important in terms of the time value of money. However, since LCC is a comparative analysis, inflation may be omitted from the LCC exercise altogether.

Since LCC deals with the future, which is uncertain, data used in LCC are also associated with uncertainty. Sensitivity analysis may be used to identify the extent to which the estimated values are sensitive to changes.

Chapter 5 examines the environmental impacts of the use of resources in construction activities, techniques of impact assessment and the application of environmental impact assessment to houses.

References

Abraham, D. M. and Dickinson, R. J. (1998) Disposal costs for environmentally regulated facilities: LCC approach. *Journal of Construction Enginering and Management* 124(2): 146–154.

Adalberth, K. (1997) Energy use during the life cycle of single unit dwellings: Example. *Building and Environment* 32(4): 321–329.

Awerbuch, S. (1993) Issues in the Valuation of PV/Renewables: Estimating the Present Value of Externality Streams with Degression on DSM. In: *Proceedings of the National Regulatory Conference on Renewable Energy*. NARUC, Washington.

Bird, B. (1987) Costs-in-use: principles in the context of building procurement, *Construction Management and Economics* 5(4): S23–S30.

De Brito, J. and Franco, B. (1994) Bridge Management Policy Using Cost Analysis. In: *Proceedings of the Institute of Civil Engineers (Structures and Buildings)*, 104 (November), pp. 431–439.

Dhillon, B. S. (1989) *Life Cycle Costing: Techniques, Models, and Applications*. Gordon & Breach Science Publishers.

Donn, M., Lee, J., Isaacs, N. and Bannister, P. (1995) Decision Support Tools for Building Code Energy Efficiency Compliance. In: *Proceedings of the International Building Performance Simulation Association Fourth International Conference, Session 3C: Building Performance*, 14–16 August, Madison, Wisconsin.

Ferry, D. J. O. and Flanagan, R., (1991) *Life Cycle Costing: A Radical Approach* (CIRIA Report No. 122). Construction Industry Research and Information Association.

Flanagan, R., Norman, G., Meadows, J. and Robinson, G. (1989) *Life Cycle Costing: Theory and Practice*. BSP Professional Books.

Gray, R. H., Bebbington, J. and Walters, D. (1993) *Accounting for the Environment*. Paul Chapman.

Gluch, P. and Baumann, H. (2004) The life cycle costing (LCC) approach: A conceptual discussion of its usefulness for environmental decision-making. *Building and Environment* 39(5): 571–580.

Haberecht, P. W. and Bennett, A. F. (1999) Experience with Durability Assessment and Performance-based Building Codes. *First Asia Pacific Conference on Harmonisation of Durability Standards and Performance Tests for Components in Buildings and Infrastructure*, 8–10 September, Bangkok, Thailand (reprint *BRANZ Conference Paper No. 72, 1999*).

Joint Centre for Land Development Studies (n.d.) *Life Cycle Costs for Architects: A Draft Design Manual*. Joint Centre for Land Development Studies, College of Estate Management, University of Reading.

Kirk, S. J. and Dellísola, A. J. (1995) *Life Cycle Costing for Design Professionals*, 2nd edn. McGraw-Hill Inc.

Ministry of Economic Development (2003) *New Zealand Energy Outlook to 2025*. Energy Modelling and Statistics Unit, Ministry of Economic development.

Mithraratne, N. and Vale, B. (2004) Life cycle analysis model for New Zealand houses. *Building and Environment* 39(4): 483–492.

Nicolini, D., Tomkins, C., Holti, R., Oldman, A. and Smalley, M. (2000) Can target costing and whole life costing be applied in the construction industry? Evidence from two case studies. *British Journal of Management* 11(4): 303–324.

Seeley, I. H. (1996) *Building Economics*, 4th edn. Macmillan Press Ltd.

Standards New Zealand (n.d.) *Life Cycle Costing: An Application Guide (Australian/New Zealand Standard)*. AS/NZS 4536:1999, Standards Australia & Standards New Zealand.

Sterner, E. (2000) Life-cycle costing and its use in the Swedish building sector. *Building Research and Information* 28(5/6): 387–393.

Stone, P. A. (1980) *Building Design Evaluation: Costs-in-Use*, 3rd edn. E. & F. N. Spon Ltd.

Wallace, W. A. (1987) Capital costs versus costs-in-use: A content analysis of design team member communication patterns. *Construction Management and Economics* 5(4): S73–S92.

5 Environmental impact assessment

Buildings and their services, while providing the occupants with a comfortable internal environment, also affect the external environment. This impact on the external environment could, in turn, affect the performance and the viability of the buildings during their useful lifetime. The environmental impact may be felt at a range of scales:

- internally – owing to the effects of building materials on the health of building workers and occupants;
- locally – due to the effects of activities such as quarrying for building materials and disposal of waste; or
- globally – as a result of the carbon dioxide emissions released due to energy used (see later Environmental impacts of building construction).

The scale of the environmental impact due to buildings depends on the decisions taken over the entire useful life of the building, including manufacturing of building materials and components, design, construction, use and maintenance, demolition and waste disposal or reuse.

Development and the ecosystem

Irrespective of technological advancement, mankind depends on the productivity and the life support services of the natural assets of the planet for basic needs and the production of material resources. The stock of natural assets comprises:

- renewable assets – resources that are self-producing and self-maintaining using solar energy and photosynthesis; e.g. living species and ecosystems;
- non-renewable assets – resources of which the supply is limited, such that any use implies liquidating part of the stock; e.g. fossil fuels and minerals; and
- replenishable assets – non-living resources that depend on solar energy for renewal; e.g. groundwater and the ozone layer (Rees 1996: p. 198).

These natural assets provide 'services' such as waste absorption, erosion and flood control, and protection from UV radiation. If these services are to operate smoothly, the ecosystem must function as an intact system. When the human-induced impact on the natural systems exceeds the equilibrium stage, unpredictable ecosystem restructuring, such as erratic climate change, becomes a possibility.

Daly (1990) has identified the following three factors as constituting the sustainable limits to resource consumption. These are:

1. rate of consumption of renewable resources that is less than the rate of regeneration;
2. rate of consumption of non-renewable resources that is less than the rate at which renewable substitutes can be generated;
3. rate of emission of pollutants that is less than the rate at which they can be recycled, absorbed or rendered harmless by the environment.

However, during the last century, the world population and the scale of human activity have both increased enormously. The growth in the world population and industrial activities is not linear but, rather, exponential. Exponential growth is a mathematical phenomenon, the rate of growth depending on the size of the existing sample. Therefore, with increasing world population and industrial activities, the corresponding growth rate also increases. Although the rate of growth varies among different countries, growth is a dominant pattern in modern society in nearly every part of the world. Developed countries consider growth as essential for employment, social mobility and the advancement of technology, while developing countries consider growth as the only way out of poverty (Meadows et al. 1992: p. 5).

Goodland and Daly argued (1996: p. 1004) that growth consists of an increase in size (amount, degree) by assimilation and accredition, whereas development is to expand, to bring out potentialities and capabilities, and to advance from a lower state to a higher state. Therefore, while development which promotes qualitative improvement might be sustainable, growth is not. As a result of exponential growth, continuously increasing flows of energy and materials are required from the environment to fulfil the requirements of the present society. However, while the human population and the rate of consumption increase daily, natural resources are in a steady state or decline.

Nonetheless, the energy and materials extracted from nature are not completely lost, as they are returned eventually in a degraded state in the form of waste and pollutants or low-grade heat. Beyond a certain level, continuous growth could lead to the depletion of natural resources as the consumption exceeds the natural regenerative capacity. This is the case with expanding demand for oil in the face of a downturn in production as oil reserves are exhausted, giving rise to so-called peak oil. Continuous unchecked growth would eventually lead to reduced biodiversity, air, water and land pollution, deforestation, etc.

Carrying capacity

Meadows et al. (1992: p. 261) defined carrying capacity as the size of the population that can be supported indefinitely by a delineated habitat. This concept was originally applied to relatively simple populations or resource systems, such as the number of cattle or sheep that can be maintained on a specified area of grazing land without permanently impairing the productivity of that land. When applied to human populations, this concept becomes more complex and irrelevant owing to the seeming ability of mankind to increase the carrying capacity through technology and trade, by importing resources that are locally scarce and by eliminating the other competing species. (As the natural resources and sinks are limited, the carrying capacity to support the human population, however, cannot be expanded indefinitely.)

As a result, Rees (1996: p. 197) redefined human carrying capacity in terms of the maximum load – and not the population – that could be safely imposed on the environment, as human beings differ from the other species due to industrial activities. Therefore, the human carrying capacity is defined as the maximum rates of resource harvesting and waste generation that can be sustained indefinitely without progressively impairing the productivity and functional integrity of relevant ecosystems, irrespective of the location of those ecosystems.

Technological advancement and innovation are generally considered to increase carrying capacity by the efficient use of resources, which would enable planet Earth to support a given population at a higher material standard or a larger population at the existing material standard. However, the gains from efficiency generally increase consumption by the provision of additional opportunities, such as cheaper fuels, lower prices and increased wages. Therefore, technological advancement indirectly reduces the carrying capacity by increased aggregate consumption. Ecological changes in the form of ozone depletion, soil loss, groundwater depletion, deforestation, loss of biodiversity, etc. are direct evidence of the fact that the aggregate consumption of human beings has exceeded the carrying capacity of the natural system in certain categories. As pointed out by Rees (1996: p. 210), the ultimate survival of a complex system that depends on a number of essential inputs and sinks, is limited by the single variable that is in shortest supply.

Environmental impact assessment

Studies on the capacity of nature to support human activities started in the late 1960s (Beck 1991). While these studies focus on various indicators such as energy requirements, non-renewable resources or even photosynthetic potentials, they are based on the principle of quantifying energy and resource flows through human society.

Limits to growth

The Club of Rome's[1] project, the Predicament of Mankind, was one of the pioneering works aimed at identifying the limits to growth in population and industrial capital. This project investigated the question of the sustainable population that can be supported by the earth together with the sustainable level of material wealth. The project used a computerised world model based on existing patterns, trends and interrelationships of the physical aspects of human society over a period of 100 years. Population, food production, industrialisation, pollution and consumption of non-renewable natural resources were the parameters modelled. Although numerical answers to the investigations were not possible owing to the simplicity of the model coupled with the data quality, information on the causes of growth and limits to growth were able to be gathered (Meadows et al. 1974: p. 94).

The model was, however, criticised for its exclusion of scientific and technological advancements that were apparent at the time (ibid.). While the physical aspects of human activities were well represented, the critical social factors that affect value systems, such as education and employment, were not included. Although population and income stabilisation could change social attitudes, the social variables included in the model were based on historical information. As the model was for the world in general, the conclusions of the study were not applicable to any particular country or region and therefore lacked the possibilities for generating any action. However, one major contribution of the report was highlighting the importance, and impossibility, of continued exponential growth in a finite world.

Human appropriation of net primary productivity

Another method of measuring the impact of humans on the earth is the amount of net primary production that is used for human activities. Net primary production (NPP) is defined by ecologists as the amount of energy captured from sunlight by green plants and fixed into living tissues (Meadows et al. 1992: p. 65). This is the basis of all food chains. Vitousek et al. (1986) argued that, although the human population directly consumes only 3% of the NPP in terms of food, animal feed and firewood, another 36% is consumed indirectly as crop waste, burning and clearing of forests, desertification and conversion of natural areas into settlements. While the latter figure does not include the negative effects of pollution, this indicates that the human population alone uses about 40% of the net photosynthetic production on land and 25% of photosynthetic production as a whole (both land and sea). As human beings take more and more NPP for their use, other life forms on earth are left with less and less, ultimately leading to their extinction.

IPAT formula

The concept of carrying capacity, which can be easily defined for other species, is inapplicable to human populations, as argued by Rees (1996), owing to the major differences that exist in terms of behaviour, technology and affluence. The maximum number of people that can be supported may not be the optimum, as both biological and industrial consumption relating to a population of people have, in turn, to be supported. The total environmental impact caused by a human population can be calculated using the IPAT formula (Ehrlich and Holdren 1971; Holdren and Ehrlich 1974; Ehrlich and Ehrlich 1990):

$$I = P \times A \times T$$

where I = the environmental impact of the population on the natural sources and sinks; P = population; A = the level of affluence; and T = the damage done by the technologies that support that affluence.

Therefore, a reduction in the impacts of human activities on the environment could only be achieved by limiting the population, by limiting affluence or by improving technology to reduce the energy and materials used for production.

Meadows et al. (1974: p. 114) argued that the per capita income affects the desired birth rate (and therefore population growth) by reason of the monetary value attached to a child. In a less well developed country, a child may be valued for his/her labour contribution to the family farm or business in addition to the eventual support of parents in their old age, while in a developed country this monetary value is reduced owing to child labour laws, compulsory education, social security provisions, etc. However, as pointed out by Meadows et al. (1992: p. 29), some countries[2] have demonstrated that a reduction in population growth is achievable without significant increase in per capita income. Factors such as education, employment for women, family planning and reduction in infant mortality play a more important role in this respect than the per capita income.

Ecological footprint calculation

As a result of advanced technology and global trade, the ecological locations of human populations no longer coincide with their geographic locations. Cities and regions of present-day society depend on the ecological productivity and life-support functions of distant localities all over the world. However, for all material and energy flows, there must be a corresponding ecosystem source or sink, and there must be biologically productive land and water to sustain these flows.

The ecological footprint concept is an area-based estimate of the natural resources and service flows required to sustain the consumption patterns of a given population, and, therefore, the amount of 'nature' used and the resulting environmental impact. It provides a useful tool that can be used to quantify the human use of nature, in order to reduce it. Calculations of ecological footprint are based on the two assumptions that:

1. it is possible to keep track of most of the resources consumed and many of the wastes generated by human activities; and
2. most of these resource and waste flows can be measured in terms of a biologically productive land area (Wackernagel et al. 1999: p. 377).

The ecological footprint represents the area of biologically productive land and water required exclusively to produce the resources consumed, and to absorb the wastes generated, by a defined population, using the prevailing technology. Therefore, it is an indication of the ecological cost of supplying goods and services to a defined population. The area of the footprint depends on the size of the population, material living standards, technology being used and ecological productivity. For most industrialised countries, the national footprint area exceeds what is available locally. This means that they run an ecological deficit. However, ecological footprints do not overlap, and therefore the global carrying capacity appropriated by the industrialised nations (or any nation) is not available to others (Wackernagel et al. 1999). Therefore, for each person that consumes three times the amount available, there are three others using only one third of the world average.

The methodology used in calculating the footprint of a given population according to Wackernagel and colleagues (1999) is as follows. As the land and sea area appropriated for production of major consumption items of any population is scattered all over the world as a result of global trade, the space required to provide the ecological services is calculated based on the world average productivity, so the footprints of various nations are comparable. Six main categories of productive areas are used to calculate the productivity. These are:

1. fossil energy land – recently forested areas with immature forests reserved for carbon dioxide absorption;
2. arable land – productive land used for cultivation;
3. pasture – grazing land for dairy and cattle farming, which is significantly less productive than arable land;
4. forest – farmed and natural forests which yield timber products;
5. built-up areas – land used for settlements, roads, etc., which generally lies in the most fertile areas; and
6. sea – which provides the marine production to supplement the human diet.

Based on the above categories, the ecological capacity that exists within a country on a per capita basis is calculated.

The calculation of the ecological capacity used is based on two main categories, consumption and energy. Consumption of a given population is considered to be the addition of production and imports, less exports. However, this consumption is termed 'apparent consumption', as this is not the actual consumption within the country. This includes the embodied resources and waste discharges of the export goods, while it does not include the same for import goods. This results in a higher footprint for a country with high exports, such as New Zealand. The energy balance of traded goods takes account of the energy directly consumed in the country and embodied energy that enters and leaves the country in terms of import and export goods. Since energy makes a significant contribution to the footprint of any country, it is analysed separately. Using annual productivity, consumption and energy are converted to land and water areas. Footprints are normally calculated on a per capita basis so that all countries are comparable. Therefore, multiplying the per capita footprint by the population gives the total footprint of a nation. According to a global study, the ecological footprint of the global population in 1997 was about 30% over the carrying capacity of the earth (Loh 2000). The underlying concept of ecosystem accounting is used in other studies, similarly to the ecological footprint but with a more limited scope, such as those by Folke et al. 1997 (resource use and waste) and Brown and Ulgiati 1998 (energy).

The ecological footprint calculation has been criticised for its methodological deficiencies (Cox et al. 2000, 2004; Pearce 2005). The basis of ecological footprint calculation is the biologically productive land and sea areas required to sustain a given population. However, the human population and the resource base are not constant, while the calculation of biologically productive land area is problematic as it involves making a judgement of the level of productivity. Further, the use of technologies can significantly increase the sustainable productivity of land, while human activities and technologies could also have negative impacts on the productivity of land.

The footprint calculation considers growing biomass as the only means of carbon sequestration, although alternative methods such as emission reduction strategies have begun to play a significant role in addressing this issue. Other proposed alternatives include capturing carbon dioxide, liquefying it and pumping it into holes in the ocean floor or oil and gas fields to replace the fuels extracted. In addition, footprint calculations do not include other important factors such as:

- uses of nature for activities such as waste absorption, pollution and industrial contamination;
- fresh water, which is a critical resource for some dry countries, where high energy costs and ecological impacts result due to water scarcity; and
- embodied resources and waste discharges of trade goods leading to higher than actual footprints for countries with high exports.

The methodology has also been criticised for measurement issues such as aggregation and substitutability (Van Kooten and Bulte 2000). Some researchers, however, argue that in spite of the methodological problems the ecological footprint calculation is valuable as a method that can provide useful qualitative insights (Costanza 2000; Moffatt 2000; Lowe 2006).

The input–output analysis-based ecological footprint calculation is an attempt to improve on some of the shortcomings of the original calculation methodology (Bicknell et al. 1998; Ferng 2001; Lenzen and Murray 2001). The basic assumption used in this method is that physical by-products of production processes such as pollution are directly tied to the economic activities used in input–output models (Bicknell et al. 1998). Since input–output data are collected in most developed countries as a part of the national accounts, these data could easily be updated. In order to calculate domestic consumption more accurately, the method uses government statistics on balance of trade. While exports are subtracted, imports are analysed using production technology similar to the domestic economy, owing to lack of data. It is assumed that all imports are final products. However, if imports are in a relatively unfinished state this assumption leads to a higher than actual footprint as some of the imported land would be supporting exports and only a fraction would support the domestic consumption.

The main advantage of this method is that it facilitates detailed analysis of the impact of international trade on the ecological footprint of a nation. With this method it is possible to determine accurately the extent of land that is being 'imported' in the form of goods and services for domestic consumption as well as the extent of land that is being 'exported' in the form of exports. Input–output-based studies have shown that New Zealand ecologically supports other economies to a greater extent than those economies support New Zealand, while living well within its own carrying capacity (Bicknell et al. 1998; McDonald and Patterson 2003).

A basic assumption used in the input–output method is that each industry produces a single product and that all output uses the same process and technology. Therefore, when footprint calculations based on this method are used for detailed analysis of a product or a process, this could lead to problems, although calculations for an entire nation are not affected. Further, the input–output method does not include non-monetary activities and therefore unpaid work is not taken into account. If a significant portion of economic activities is of this type, the use of the input–output method to calculate ecological footprint may not be appropriate (Bicknell et al. 1998).

For sustainability, the average footprint needs to be reduced to the global ecological capacity. If one person uses more than this, to compensate others have to use less. If the footprint exceeds the area available within the country for biological productivity, the country runs an ecological deficit, as the ecological services used by the population cannot be provided within it. Therefore, the country needs either to import the missing capacity or to deplete its local natural resources. As a result of high biological productivity, some countries may be able to support their citizens at a higher level of resource consumption than found in the rest of the world. Often such countries use the remaining ecological capacity to produce export goods rather than reserving it. Global ecological deficit is the gap between the average consumption of the region where a person is living and the capacity available per person in the world (Wackernagel et al. 1999: p. 385). The footprint calculation, although designed for the individual, seems to be a possible method of evaluating the environmental impact of a building, as buildings are demanded by and used by individuals, although the data required for such a calculation would be numerous.

Environmental impacts of building construction

Like other human activities, building construction also contributes to environmental impacts, which can either be:

- short-term – e.g. noise and dust; or
- permanent – e.g. carbon dioxide emissions.

The environmental impacts of a building are broadly illustrated in Figure 5.1.

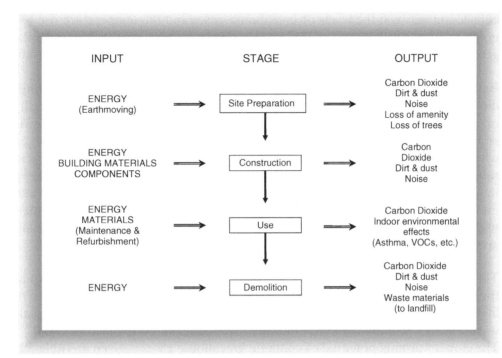

Fig. 5.1 Environmental impacts of buildings throughout useful lifetime.
Based on: Harris, D. J. (1999) A quantitative approach to the assessment of the environmental impacts of building materials. *Building & Environment* 34: 752.

The main environmental issues related to buildings are:

- global atmospheric pollution – greenhouse effect, acid rain, ozone depletion, etc.;
- depletion of resources and effects on the local environment – use of natural materials for buildings, impacts on wildlife, etc.;
- health, comfort and safety of occupants – indoor environment; and
- impact of climate change on buildings – change in temperature, wind loads, driving rain, etc. (Prior et al. 1991).

Global atmospheric pollution

The greenhouse effect is caused by absorption of infra-red radiation, emitted by the earth's surface, by certain greenhouse gases. Although this is a natural phenomenon, human activities have increased the greenhouse effect by increasing the atmospheric levels of gases such as carbon dioxide, methane, oxides of nitrogen, ozone and chlorofluorocarbons (CFCs) above those of the past. There is evidence that global temperatures have been rising, although the precise connection between the levels of greenhouse gases and the temperature rise has not been established. However, if there is a connection between the two, and if the greenhouse gas levels continue to rise, major climate changes might be the result.

Buildings contribute to the formation of gases such as carbon dioxide, sulphur dioxide and oxides of nitrogen through the combustion of fossil fuels, especially oil and coal, whether directly or to generate electricity. While carbon dioxide emissions increase the greenhouse effect, sulphur dioxide and oxides of nitrogen when dissolved in rainwater form diluted sulphuric and nitric acids to create acid rain (Bishop et al. 1995: p. 116). Acid rain can damage rivers and lakes and has a long-term effect on fresh water and soil, and could also affect forests, crops and building materials. Greenhouse gases are emitted by buildings throughout their useful life, although, in terms of quantity, carbon dioxide is the predominant greenhouse gas for New Zealand houses (Camilleri 2000a: p. 55).

The impact of different greenhouse gases in terms of global warming is known as the global warming potential (GWP). As different greenhouse gases have different lifetimes in the atmosphere, global warming potential varies with the timescale used. The most commonly used timescale is 100 years and is denoted as GWP^{100}. The global warming potential of common greenhouse gases is given in Table 5.1.

Electricity is generated in New Zealand mainly using hydro, thermal and geo-thermal power stations providing 64%, 26% and 6% of total requirements respectively. Biogas, waste heat, wood and wind supply the balance (MED 2005b: p. 120). Greenhouse gas emissions due to electricity generation in New Zealand for the year 2004 are given in Table 5.2.

Table 5.1 Global warming potential of common greenhouse gases relative to carbon dioxide

Greenhouse gas	Global warming potential (GWP^{100})
Carbon dioxide (CO_2)	1
Methane (CH_4)	21
Nitrous oxide (N_2O)	310
Hydrochlorofluorocarbons (HFCs)	140–11,700
Perfluoromethane (CF_4)	6,500
Perfluoroethane (C_2F_6)	9,200
Sulphur hexafluoride (SF_6)	23,900

Based on: Taylor, R. and Smith, I. (1997) *The State of New Zealand's Environment 1997.* Ministry for the Environment, p. 5.29

Table 5.2 Greenhouse gas emissions due to electricity generation in New Zealand (2004)

Greenhouse gas	Emissions factor (t/GWh)
Carbon dioxide (CO_2)	145.1
Methane (CH_4)	0.0036
Nitrous oxide (N_2O)	0.0017
Oxides of nitrogen (NO_x)	0.5670
Carbon monoxide (CO)	0.0390
Non-methane volatile organic compounds (NMVOCs)	0.0100

Based on: Ministry of Economic Development (2005a) *New Zealand Energy Greenhouse Gas Emissions 1990–2004*. Ministry of Economic Development (2006); and Ministry of Economic Development (2006) *New Zealand Energy in Brief – March 2006*. Ministry of Economic Development.

The table shows that carbon dioxide is the predominant greenhouse gas emission for electricity generation in New Zealand. Since hydro power stations currently meet only around 64% of the total electricity demand in New Zealand, any increase in demand has to be met by thermal stations with higher carbon dioxide emissions. Further, almost all of the cheap and easy hydroelectric sites have been developed and it appears that most of the new power stations are thermal (gas and coal), and therefore every new building calls into existence part of a new thermal power station to meet its demand. The emissions factor for average electricity is $0.1\,kg\,CO_2/kWh$, while it is $0.64\,kg\,CO_2/kWh$ for thermal electricity (Camilleri 2000a).

Common greenhouse gases emitted by New Zealand houses are carbon dioxide, nitrous oxide, methane, CFCs and perfluorocarbons (PFCs) (ibid.). Jaques et al. (1997), cited by Camilleri (ibid.), found that operating energy constitutes 76% of the life cycle energy of the average New Zealand house (see also pp. 163–164). Electricity is the predominant energy source used in New Zealand houses (see Table 3.3). Therefore, carbon dioxide is the main greenhouse gas emission from New Zealand houses. Further, significant amounts of carbon dioxide are also emitted during the manufacture of building materials. While carbon dioxide is released due to process energy requirements, additional amounts are released during manufacture of concrete, lime, steel and aluminium (Honey and Buchanan 1992: p. 29). In addition to increased carbon dioxide emissions, aluminium refining emits PFCs, which have a very high global warming potential, ranging from 6,500 to 9,200, and sulphur dioxide, which could cause acid rain. Greenhouse gas emission factors due to chemical reactions taking place during the manufacture of the main building materials used in New Zealand are given in Table 5.3.

Table 5.3   Greenhouse gas emission factors (t/t) due to chemical reactions during the manufacture of the main building materials used in New Zealand

Material	CO_2	CO	NO_x	SO_2	PFCs
Cement	0.40	0.0000	0.0000	0.00065	–
Lime	0.73	–	–	0.00048	–
Iron and steel	1.94	0.0011	0.0011	–	–
Aluminium	1.58	0.1091	0.0020	0.02000	0.2304

Based on: Ministry for the Environment (2006) *New Zealand's Greenhouse Gas Inventory 1990–2004*. Ministry for the Environment, pp. 15, 146–147

Nitrous oxide is mainly emitted by petrol and diesel vehicles used during building construction and demolition phases and by fossil fuel boilers used for water heating in some New Zealand houses. However, these emissions are negligible compared with carbon dioxide emissions. Methane emissions in the construction sector are due to transport and manufacturing processes using fossil fuel and thermal electricity. However, the main methane emissions in New Zealand are due to livestock farming. Although the main purpose of sheep farming in New Zealand is meat production, wool is a by-product which provides a significant income. Therefore, part of the methane emissions due to sheep farming could be attributable to wool production, apportioned based on the income generated. This means that insulation products based on wool contribute to global warming, particularly since the GWP^{100} of methane is 21.

Chlorofluorocarbons, hydrochlorofluorocarbons (HCFCs) and halons have been identified as causing damage to the ozone layer in addition to their global warming potential. These were used extensively in air-conditioning systems and fire extinguishers used in buildings during the 1970s. Owing to their lack of chemical reactivity, these substances can remain unchanged for many decades in the atmosphere, causing depletion of the ozone layer. Depletion of the ozone layer could increase the level of ultraviolet radiation from the sun that is reaching the earth. Increased levels of ultraviolet radiation are linked with human sunburn, skin cancer and eye cataracts, reduced crop yields, damage to trees and aquatic organisms, and increased rates of degradation of building materials (Bishop et al. 1995: p. 115). Although CFCs have been used as a blowing agent for insulation, the manufacture of New Zealand-made expanded polystyrene products has not involved the use of CFCs or HCFCs, but pentane (Camilleri 2000a: p. 54). However, it is still not possible to balance directly the environmental impact (global warming potential) of wool, a naturally occurring insulator, against the non-ozone-damaging insulation products that use oil, a diminishing resource, as a starting point for their manufacture. To some extent, all materials will make an impact on the environment.

Local environmental effects

Buildings use a variety of materials for their construction, and extraction of some of these could create local environmental effects. Production of building materials uses varying amounts of energy, as some materials need simple processing while others use energy intensive processes. Some building materials potentially can be recycled to reduce environmental effects. Recycling follows two types of processes, the outcome depending on which of these processes is used. Materials recycled using the closed loop method can be employed for the same purpose after recycling – steel is an example of such a material. However, if the open loop method is used due to the inability to separate materials during the recycling process, recycled materials cannot be reused for the same purpose – concrete is an example, as broken-up concrete can only be used as the aggregate for new concrete, with the addition of fresh cement and water (Polster et al. 1996: p. 222).

However, the actual saving in energy and, therefore, the reduction in the environmental impact due to the use of recycled building materials, depends on the quantity of the material used in the building and its embodied energy. Harris and Elliot (1997), cited by Harris (1999: p. 754), argued that the viability of a recycling process also depends on these factors. According to Harris (ibid.), for high-density, low-value materials such as concrete, transport energy constitutes a major portion of the embodied energy, and therefore the viability of recycling depends on the distances involved. For a high-value, low-density material, recycling is almost always viable as the embodied energy of the recycled product is generally about one tenth of that of the virgin material. The maximum distances a recycled material can be transported such that the environmental impact is less than for new materials locally manufactured, according to the Building Research Establishment of the UK, are given in Table 5.4. However, these figures are applicable only to the UK situations, as embodied energy coefficients vary considerably between countries (see Chapter 3, Embodied energy). For example, the UK imports 90% of its timber, leading to a higher initial embodied energy, which may explain the long distance that recycled timber can be transported with less environmental impact than that created by locally manufactured new timber.

While recycled concrete can be used as aggregate for new concrete intended for lower-quality end-products such as residential driveways, use of scrap as 25% of the raw material in steel manufacture can reduce energy use by 50% (Honey and Buchanan 1992).

Table 5.4 Maximum transport distances for recycled materials

Material	Distance (miles)
Tiles	100
Slates	300
Bricks	250
Aggregate	150
Timber (e.g. floorboards)	1,000
Steel products	2,500
Aluminium products	7,500

Source: Anderson, J. and Howard, N. (2000) *The Green Guide to Housing Specification: An environmental profiling system for building materials and components.* Construction Research Communications Ltd. (for Building Research Establishment Ltd.), p. 5

Factors governing recycling and reuse of materials according to a New Zealand study (Synergy Applied Research Limited and Department of Trade and Industry 1985) are:

- value of virgin material;
- cost of transport – distance of the reprocessing facility and density of material;
- condition of the waste material;
- nature of source separation;
- demand for recycled material;
- volume of waste materials available; and
- government support and the public involvement in recovery of materials.

Impact of climate change on buildings

Since buildings have a long life, any change of climate could have adverse implications on the buildings and the current design practices. The most important potential impacts on New Zealand houses have been identified as greenhouse gas emissions, overheating, and coastal and inland flooding (Camilleri 2000b).

Environmental impact assessment of buildings

As determined by Harris (1999: p. 751), the characteristics of a building environmental assessment system are:

- knowledge of the different ways by which the environment is affected;
- a measure of the scale of environmental effects; and
- a yardstick against which to measure these effects.

Building environmental assessment systems are the medium for creating interest in and focus on building environmental issues. Further, assessment systems promote higher environmental expectations and influence building performance by providing a means to demonstrate such performance. They also encourage the discussion of building performance by offering a simple structure for the complex issues involved. The initial objective of building environmental assessment was to provide an objective evaluation of the resource use, ecological loadings and indoor environmental quality of buildings (Cole 2005) so that buildings with a higher performance could be identified.

There are many environmental assessment schemes for buildings in use internationally, with issues being assessed at various depths. Since the introduction of the Building Research Establishment Environmental Assessment Method (BREEAM) in 1990, a number of techniques that can be used by practitioners in various contexts and life stages of buildings have been developed. Notable techniques include Leadership in Energy and Environmental Design (LEED), the National Australian Built Environment Rating System (NABERS), Comprehensive Assessment System for Building Environmental Efficiency (CASBEE), Comprehensive Environmental Performance Assessment Scheme (CEPAS), Ecoeffect and GBTool. Some of these schemes are briefly discussed below.

BREEAM has now developed assessment methods for new homes, new offices, existing offices, superstores and supermarkets. The Building Environmental Performance Assessment Criteria (BEPAC) of Canada, Hong Kong Building Environmental Assessment Method (HK-BEAM) and the BRANZ Green Home Scheme of New Zealand are based on the BREEAM scheme. As these schemes were originally developed as an introduction to more sustainable building practices, they focus on the assessment of a building at the design stage and not on the total life of the building. Since the effects considered by the above schemes cover a wide range, comparisons between different effects are not possible, and the points awarded for various categories are totalled to give a single figure. However, as argued by Harris (1999: p. 752), keeping the individual effects separate in the assessment may be beneficial, as an improvement in the performance of one aspect might lead to the deterioration of another. (For example, the space heating energy requirement may be reduced by increasing thermal mass but this would also increase the embodied energy.) Further, such a scheme allows competing designs to be compared based on the significant indicators, and the designs could be modified to improve the identified indicators.

NABERS[3] is an annual rating scheme for all building types existing and new, and therefore provides a picture of the actual performance of buildings in use. It was intended as a basic environmental impact rating of the national building stock of Australia which could be used to evaluate the trends in building activities such as use of energy in buildings, useful life, floor area of buildings, etc. Further, annual assessment could reveal equipment failure and need for maintenance. Currently, NABERS is available for commercial office buildings in both a tenant and an owner version. NABERS assessment is based on a simple spreadsheet that can be filled out by the building owner/user, based on information such as utility bills, although a certified assessor must be employed to obtain an official rating. The annual rating provides an indication of the current status of the building stock with respect to criteria for environmental sustainability.

Both ATHENA (Cole et al. 1996) and The Office Toolkit (Bishop et al. 1995) are assessment methods that employ life cycle analysis. ATHENA covers most building types and can be used to compare design alternatives while Toolkit is intended for office buildings. Toolkit measures the relative importance of the different environmental impacts in terms of 'eco points'. The eco point is a measure of the environmental damage caused and is based on a society's desire to reduce such damage (Bishop et al. 1995: p. 2). The method is used to highlight priority areas, but does not quantify the environmental impacts (Bishop et al. 1995: p. 3).

The Green Guide by the Building Research Establishment in the UK facilitates the comparison of environmental performance of competing construction types based on common specifications used for residential building elements in the UK. The method uses life cycle assessment data derived for the UK manufacturing industry. The designer can select specifications from a range of available alternatives. In order to make meaningful comparisons, functional units have been defined for each category. Ratings compare $1\,m^2$ of specification over a 60-year building life, with maintenance, refurbishment and demolition aspects taken into consideration. Environmental issues considered in the Green Guide reflect the concerns related to production and use of building materials in the UK. Different impacts are individually rated together with an overall rating. The individual ratings allow the designer to select the specifications based on the performance against a particular environmental parameter, while overall rating enables the selection to be made based on the overall performance (Anderson and Howard 2000: p. 1).

The external walls, upper floors and roof of a typical UK house account for more than 50% of the total mass, while substructure and ground floor take up another 20% (Anderson and Howard 2000: p. 6). Owing to the high mass of these elements, they have the potential to make the greatest environmental impact and, therefore, are considered to need particular attention from the designer.

However, as many buildings last longer than 60 years, the value of low maintenance and long-lasting buildings is underestimated by the rating while temporary buildings, with less mass and, hence, lower initial embodied energy, but that have a short life, are overestimated. Although the method includes life cycle impacts of material use, operating requirements other than maintenance are not considered.

Assessment schemes developed in individual countries suffer from their unsuitability for comparing results between countries. GBTool is a common method used for performance evaluation of a range of buildings in countries participating in the Green Building Challenge. To address variations in region specific conditions, it uses a generic framework, methodology and terminology with regional variations in technical and social aspects (Larsson and Cole 2001).

A computer model by Polster et al. (1996) for comparison of alternative designs using life cycle analysis considers only the external impacts. Although Polster and colleagues have argued that the products compared using life cycle analysis should fulfil the same function, which is correct, the use of a unit area as the functional unit in environmental impact assessment is questionable. Generally, when two buildings are compared on a unit area basis, the larger building tends to perform better, although the overall impact is greater. Hence, the use of additional energy and materials and the resultant higher impact from constructing the larger building are not clear in such a comparison. Further, this method uses average aggregate values on a global basis, although the actual environmental impact is more location specific.

Harris (1999) devised an assessment scheme based on the quantity of material used in a building, with various indicators kept separate, and argues it to be quantitative. However, the scale used for some of the indicators such as scarcity of raw materials is subjective and is therefore of limited value. Further, the scheme is focused on the design stage and does not include life cycle assessment. The environmental impact of a building depends not only on the actions at the design stage but on those throughout the life of the building. The selection of the building site, design, user behaviour, etc., all contribute to this impact.

The Green Home Scheme developed by the Building Research Association of New Zealand (BRANZ 1997) is a rating scheme for new residential construction, aimed at providing designers with a tool which will allow a building to be assessed at the design stage, so that various design options and strategies can be compared with one another based on the performance over their useful lifetime. However, the Green Home Scheme approaches environmental impact rating using a broad-brush manner that considers a wide range of criteria.

Cole (2005) divided building assessment techniques into two distinct types: tools and methods. A tool is defined as an assessment technique which estimates one or more environmental performance characteristics, such as embodied and operating energy use, and greenhouse gas emissions. The main distinction between the various tools available is whether or not a tool is based on life cycle analysis. While life cycle analysis-based tools evaluate the resource use and the design options, non-life cycle analysis tools rate building performance through rating/weighting systems (using aggregation of points and expert panel consensus). The use of tools is generally voluntary (e.g. ATHENA; Envest). A method also involves the use of assessment, but may need third party confirmation to achieve a rating or weighting. Generally, methods are managed and operated in an organisational context (e.g. BREEAM; LEED).

Over the years the focus of building assessment has shifted from evaluation of a building's performance (physical design and functional performance) against typical or predefined performance criteria to sustainable building assessment which addresses broader environmental, social and cultural issues related to buildings. Kaatz et al. (2006: p. 310) differentiate the two approaches as green assessment techniques and sustainable assessment techniques. Green assessment techniques (e.g. BREEAM; LEED) focus on the physical building and generally rely on relative assessment, while sustainable techniques (e.g. SBAT – Sustainable Building Assessment Tool; SPeAR – Sustainable Project Appraisal Routine) focus on the building process and measure the distance-to-targets pre-established by the client and the design team (ibid.).

Kaatz et al. (2006) also highlight the emerging role of building environmental assessment with the shift in the assessment emphasis from appraisals of designs to those assessments which promote sustainable construction practices. They identified integration, transparency and accessibility, and collaborative learning as the main qualities of such assessment criteria. While the current assessment emphasis on improving the physical performance of a single building was questioned, they suggest that future assessments should be based on broader sustainability issues such as inter- and intra-generational equity and preserving the earth's carrying capacity (Kaatz et al.: p. 315). Equity is to be safeguarded by wider participation of building professionals and lay participants in collaborative learning during the building process, which can in turn foster positive changes in the participants' attitudes towards sustainability[4]. Carrying capacity (and ecological footprint) as a means to limit environmental impacts and achieving sustainability, however, are not accepted universally owing to the impact of international trade and other methodological deficiencies discussed earlier (see Carrying capacity and Ecological footprint calculation). Time, effort and cost involved in assessment are further issues to be considered in future developments in assessment. The checklist approach followed by a comprehensive assessment which covers a minimum number of criteria of significance at a later stage has been suggested as a future direction. Liu et al. (2006) have argued that the current assessment techniques mix performance and design related factors in the assessment framework, causing confusion. They highlight the importance of a framework which classifies assessment tools/methods in order to facilitate accurate selection of tools appropriate for a specific application.

Assessment of climate change impacts on buildings

The above assessment schemes concentrate on the effects of building construction on climate change; none considers the effects of climate change on the buildings. The Climate Change Sustainability Index (Camilleri 2000b) for New Zealand houses considers both:

- effects of climate change on New Zealand houses and their vulnerability; and
- contribution of houses to climate change as a result of greenhouse gas emissions.

However, the main objective is to assess effects of climate change on a building. As information on the greenhouse gas emissions attributable to manufacture of building materials/components was not available at the time of the development of the index, embodied energy and embodied emissions are not included. Operating requirements for space heating and water heating are included, though lighting is regarded as negligible. The other components not included are transport energy and maintenance energy.

Emissions due to space heating are quantified based on the thermal performance of a house and the heating system used on a per person per year basis (ibid.). It is argued that quantifying emissions on a per person basis would provide a comparable unit for houses of different sizes. However, a larger house that would have a greater environmental impact in terms of its embodied energy would be ignored in this analysis. Emissions resulting from hot water use are calculated based on the hot water heater type (ibid.). The normal power demand for water heaters generally occurs at peak hours, during which time the additional demand is met by thermal stations, and therefore electrical power used is assumed to be thermal. Even though the use of wood as a fuel is significant in New Zealand (Isaacs et al. 2006), it has not been included for either space or water heating. The two types of emissions discussed above are combined to derive a single rating for the impact of the building on climate change.

The level of overheating experienced by the house is assessed based on the maximum summer indoor air temperature calculated using:

- level of insulation;
- thermal mass area; and
- solar window area – area of windows facing SW through N to due E (Camilleri 2000b: p. 18).

The maximum temperature thus calculated is modified to account for the level of insulation and the location of a particular house in New Zealand.

According to Camilleri (2000a: p. 23), inland flooding annual exceedence probability (AEP) is likely to double by the year 2030, and quadruple by the year 2070. The change in coastal flooding depends on the individual site, and is not known for the majority of sites in New Zealand. Although the risk of changes cannot be predicted accurately, the potential for damage to houses by tropical cyclones is considered to be the highest. However, in rating the effects of climate change on New Zealand houses, overheating, flooding and tropical cyclones are considered to be of equal importance (ibid.).

At present, the environmental assessment schemes that are available focus on a wide range of issues and aspects to various depths. However, most assessment schemes do not encourage improvements in environmental performance of buildings as good ratings can be obtained merely by conforming to the existing building regulations. An assessment scheme, while focusing on the issues during the complete useful life of a building, should encourage better environmental performance than is possible under the current practices if the human impact on natural systems is to be reduced. However, the lack of information on the environmental issues related to production and use of building materials commonly used in the New Zealand other than for greenhouse gas emissions is a concern, which needs attention.

Conclusions

Exponential growth of human population and industrial activities is imposing increasing pressures on natural ecosystems. Many studies have been carried out to quantify the energy and resource flows through human society in an attempt to reduce the human impact. However, the impact of human activities on natural ecosystems is evident in the form of erratic climate change.

Buildings contribute to environmental impacts as a result of decisions taken throughout their useful life, which can be either temporary or permanent. Although many researchers use energy as an indication of the environmental impacts due to buildings, energy alone does not provide a complete evaluation.

The main environmental issues related to buildings consist of global atmospheric pollution, depletion of resources, health and comfort of users, and impacts of climate change on buildings.

There are numerous impact assessment schemes currently being used internationally. The method of assessment used and the system boundary vary widely between these schemes. While assessment schemes vary from building evaluation tools to rating methods, some employ the principles of life cycle analysis. An ideal environmental assessment scheme for houses should encourage better environmental performance and should cover the aspects related to total useful life. Although it would be possible to calculate the ecological footprint of a building and thereby evaluate the environmental impact of, say, a house, the amount and the nature of data required for such an assessment are not readily available, while the methodology is also fraught with defects.

The energy industry contributes a significant portion to the total greenhouse gas emissions of any economy and, therefore, unlike other building-related environmental impacts, greenhouse gases are location-specific. Carbon dioxide is the predominant greenhouse gas attributable to New Zealand houses, although limited amounts of carbon monoxide, oxides of nitrogen, sulphur dioxide and perfluorocarbons are emitted in the manufacture of the main building materials used. Although information on greenhouse gas emissions is available, lack of data on the other environmental issues related to manufacture and use of building materials commonly found in houses is a hindrance to proper assessment.

In this section, life cycle energy, cost and environmental impacts due to the building industry have been discussed in detail. The development of the life cycle analysis model and its use for evaluation of New Zealand houses are discussed in the next section.

Notes

[1] The Club of Rome is an international group of scientists, humanists, industrialists, statesmen and businessmen who are interested in the future of mankind in a continually growing society.
[2] China, Sri Lanka, Costa Rica, Singapore, Thailand, Malaysia, etc.
[3] http://www.nabers.com.au/faqs.aspx
[4] However, it is the physical performance which in the end causes the environmental damage, not the attitudes of the designers and users.

References

Anderson, J. and Howard, N. (2000) *The Green Guide to Housing Specification: An environmental profiling system for building materials and components.* Construction Research Communications Ltd. (for Building Research Establishment Ltd.).

Beck, M. B. (1991) Forecasting environmental change. *Journal of Forecasting* 10(1–2): 3–19.

Bicknell, K. B., Ball, R. J., Cullen, R. and Bigsby, H. R. (1998) New methodology for the ecological footprint with an application to the New Zealand economy. *Ecological Economics* 27(2): 149–160.

Bishop, T., Durrant, H. and Bartlett, P. (1995) *The Office Toolkit: The guide for facilities and office managers for reducing costs and environmental impact.* Building Research Establishment.

Building Research Association of New Zealand (1997) *Green Home Scheme: Home Owners' Guide.* Building Research Association of New Zealand.

Brown, M. T. and Ulgiati, S. (1998) Emergy evaluation of the environment: Quantitative perspectives on ecological footprints. In: *Advances in Energy Studies: Energy Flows in Ecology and Economy*, pp. 223–224. Museum of Science and Scientific Information.

Camilleri, M. J. (2000a) *Implications of Climate Change for the Construction Sector: Houses* (BRANZ Study Report No. 94). Building Research Association of New Zealand.

Camilleri, M. J. (2000b) *A Draft Climate Change Sustainability Index for Houses* (BRANZ Study Report No. 95). Building Research Association of New Zealand.

Cole, R. et al. (1996) *ATHENA, An Environmental Assessment of Building Designs.* Forintek Canada Corporation.

Cole, R. J. (2005) Building environmental assessment methods: Redefining intentions and roles. *Building Research and Information* 35(5): 455–467.

Costanza R. (2000) The dynamics of the ecological footprint concept. *Ecological Economics* 32(3): 341–345.

Cox, P. M., Betts, R. A., Jones, C. D. et al. (2000) Accelaration of global warming due to carbon-cycle feedbacks in a coupled climate model. *Nature* 408(6809): 184–187.

Cox, P. M., Betts, R. A., Collins, M. et al. (2004) Amazon dieback under climate-carbon cycle projections for the 21st century. *Theoretical and Applied Climatology* 78(1–3): 137–156.

Daly, H. (1990) Toward some operational principles of sustainable development. *Ecological Economics* 2: 1–6.

Ehrlich, P. R. and Holdren, J. P. (1971) Impact of population growth. *Science* 171: 1212–1217.

Ehrlich, P. R. and Ehrlich, A. H. (1990) *The Population Explosion.* Simon and Schuster.

Ferng, J. J. (2001) Using composition of land multiplier to estimate ecological footprints associated with production activity. *Ecological Economics* 37(2): 159–172.

Folke, C., Jansson, A., Larsson, J. et al. (1997) Ecosystem appropriation by cities. *Ambio* 26: 167–172.

Goodland, R. and Daly, H. (1996) Environmental sustainability: Universal and non-negotiable. *Ecological Applications* 6(4): 1002–1017.

Harris, D. J. and Elliot, C. J. (1997) Energy Accounting for Recycled Building Components. *Second CIB Conference on Buildings and the Environment*, Paris, June 1997.

Harris, D. J. (1999) A quantitative approach to the assessment of the environmental impact of building materials. *Building and Environment* 34(6): 751–758.

Holdren, J. P. and Ehrlich, P. R. (1974) Human population and the global environment. *American Scientist* 62(3): 282–292.

Honey, B. G. and Buchanan, A. H. (1992) *Environmental Impacts of the New Zealand Building Industry* (Research Report 92/2). Department of Civil Engineering, University of Canterbury, New Zealand.

Isaacs, N., Camilleri, M., French, L. et al. (2006) *Energy use in New Zealand Households*, HEEP Year 10 Report, Building Research Association of New Zealand.

Jaques, R., Bennett, A., Sharman, W. and Isaacs, N. (1997) Legislative Opportunities to Reduce CO_2 Emissions in New Zealand. *Second International Conference Buildings and the Environment*, Wellington, 1997.

Kaatz, E., Root, D. S., Bowen, P. A. and Hill, R. C. (2006) Advancing key outcomes of sustainability building assessment. *Building Research and Information* 34(4): 308–320.

Larsson, N. K. and Cole, R. J. (2001) Green Building Challenge: the development of an idea. *Building Research and Information* 29(5): 336–345.

Lenzen, M. and Murray, S. A. (2001) A modified ecological footprint method and its application to Australia. *Ecological Economics* 37(2): 229–255.

Liu, Y., Prasad, D., Li, J. et al. (2006) Developing regionally specific environmental building tools for China. *Building Research & Information* 34(4): 372–386.

Loh, J. (2000) *Living Planet Report 2000*. World Wide Fund for Nature.

Lowe, R. (2006) Defining absolute environmental limits for the built environment. *Building Research & Information* 34(4): 405–415.

McDonald, G. and Patterson, M. (2003) *Ecological Footprints of New Zealand and its Regions*. Ministry for Economic Development.

Meadows, D. H., Meadows, D. L., Randers, J. and Behrens, W. W. III (1974) *The Limits to Growth: A Report for the Club of Rome's Project on the Predicament of Mankind*. Pan Books Ltd. [First published 1972.]

Meadows, D. H., Meadows, D. L. and Randers, J. (1992) *Beyond the Limits: Global Collapse or a Sustainable Future*. Earthscan Publications Ltd.

Ministry for the Environment (2006) *New Zealand's Greenhouse Gas Inventory 1990–2004*. Ministry for the Environment.

Ministry of Economic Development (2005a) *New Zealand Energy Greenhouse Gas Emissions 1990–2004*. Ministry of Economic Development.

Ministry of Economic Development (2005b) *New Zealand Energy Data File – July 2005* (compiled by Dang, D. T. H.). Energy Modelling and Statistics Unit, Energy and Resources division, Ministry of Economic Development.

Ministry of Economic Development (2006) *New Zealand Energy in Brief – March 2006* (compiled by Dang, D. T. H. and Cowie, P.). Energy Information and Modelling Group, Ministry of Economic Development.

Moffatt I. (2000) Ecological footprints and sustainable development. *Ecological Economics* 32(3): 359–362.

Pearce, D. (2005) Do we understand sustainable development? *Building Research & Information* 33(5): 481–483.

Polster, B., Peuportier, B., Sommereux, I. B. et al. (1996) Evaluation of the environmental quality of buildings towards a more environmentally conscious design. *Solar Energy* 57(3): 219–230.

Prior, J. J., Raw, G. J. and Charlesworth, J. L. (1991) *BREEAM New Homes, Version 3/91: An Environmental Assessment for New Homes*. Building Research Establishment.

Rees, W. E. (1996) Revisiting carrying capacity: Area-based indicators of sustainability. *Population and Environment: A Journal of Interdisciplinary Studies* 17(3): 195–215.

Synergy Applied Research Limited and Department of Trade and Industry (1985) Estimated New Zealand Waste Quantities: Glass, Paper, Ferrous Metal and Plastic (Resource Conservation Series 19). Department of Trade and Industry, New Zealand.

Taylor, R. and Smith, I. (1997) *The State of New Zealand's Environment 1997*. Ministry for the Environment.

Van Kooten, C. and Bulte, E. (2000) *The Economics of Nature*. Blackwell.

Vitousek, P., Ehrlich, P., Ehrlich, A. and Matson, P. (1986) Human appropriation of the products of photosynthesis. *BioScience* 36: 368–374.

Wackernagel, M., Onisto, L., Bello, P. et al. (1999) National natural capital accounting with the ecological footprint concept. *Ecological Economics* 29: 375–390.

Part B:

Life Cycle Analysis Model, Results and Lessons

6 Development of a New Zealand model

The literature on life cycle energy, cost and environmental impacts has been surveyed in the previous section. Methods for analysis of life cycle energy and cost have been shown to be well developed, albeit with the practical problems of data quality and methodological imperfections. Although numerous environmental impact assessment schemes exist, analysis of the environmental impacts related to buildings, other than greenhouse gas emissions, is still being developed.

Throughout the useful life of a building, resources are expended and costs are incurred to maintain and operate it at a habitable level. Hence, if a building is to be more sustainable, in addition to using less non-renewable energy for operation it also has to be constructed of durable materials so that it will last longer. Although globally many designers who are concentrating on minimising the impact their buildings make on the environment are claiming that their buildings are sustainable, for a holistic evaluation of the impact a building makes on the environment, an objective analysis is required. While such an analysis should consider both the operating as well as construction requirements of various buildings, the evaluation should also cover the total useful life of such buildings. However, performance analyses in New Zealand so far have been limited to a very short useful life of 25–50 years (Buchanan and Honey 1994; Jaques 1996) based on the requirement of 50 years specified for the life of structural members, in the New Zealand Building Code (Building Industry Authority 1998). The use of this shorter lifetime undermines the potential longer-term benefits for the analysis of considering the energy embodied in the building materials used for the construction, and therefore does not represent a true evaluation of the environmental impact a building makes. Although Johnstone (2001) used 90 years for his life cycle study, some of the data used are not representative of general practice in New Zealand (see Replacement cycles below for a discussion). Furthermore, none of these studies has included the operating or embodied energy requirements of appliances and equipment or the embodied energy of furniture. All of these items require frequent replacement owing to their short useful life and therefore they could make a significant contribution to the total life cycle environmental impact.

Life cycle energy, if quantified in terms of primary energy, directly reflects the greenhouse gas emissions attributable to houses and therefore their environmental impact. However, design decisions are evaluated by individual house-owners based on the value provided for the money they spend, among many other things, and therefore the initial cost and, more importantly, the life cycle cost of design decisions, often become deciding factors.

Buildings last a long time compared with building fittings and equipment, making such an analysis a time-consuming and tedious task. It is therefore not practical for a designer to predict the effect a certain design decision would have on the environmental impact of a building over its life. It is even harder to compare one design with another. It is often useful for a designer to have a tool which will allow a building to be assessed at the design stage, so that various design options and strategies can be compared with one another based on the performance over their useful lifetimes. Further, a tool which keeps individual effects separate allows competing designs to be compared based on the indicators which are considered significant, and the designs can then be modified to improve them to reduce the impact of the identified indicators.

A life cycle analysis model can therefore facilitate a more detailed impact analysis and can aid design decisions for those designing and specifying individual residential buildings. Such models, however, should not be used to predict the life cycle performance of a particular design, as the predicted performance will seldom be matched by the actual performance, because of the influence of the users of the building, in the same way that the habits of different drivers affect the petrol consumption of apparently similar cars. However, if used for comparative analysis of alternative designs, and for assessing possible improvements to a design by modifications to and replacement of construction types used, such analysis could provide useful information during the design phase.

A Life cycle analysis model for New Zealand houses

The poor quality of data available at present is a major drawback for the development of detailed life cycle analysis-based models for building evaluations. The use of expert systems, however, which in the past has proved successful (Geraghty 1983; Culaba and Purvis 1999) for environmental evaluations, can overcome this situation. A life cycle analysis-based model representative of NZ practices was generated as a stand-alone application. It consists of three basic independent components: (1) knowledge base, (2) inference engine and (3) graphical user interface. This layout allows the model to be updated with reasonable ease as better-quality data become available and to be adapted easily even for situations in other locations.

The knowledge base contains the qualitative and quantitative data. These include:

- generic construction types based on the elements of a house;
- embodied energy of NZ building materials, appliances and furniture;
- replacement cycles for building materials/components, appliances and furniture;
- installed prices of building materials/components and prices of appliances, furniture and energy;
- operating energy requirements (energy and CO_2 emissions) for space conditioning, appliances, lighting, hot-water system, etc.;
- greenhouse gas emissions for NZ building materials, appliances and furniture; and
- environmental impact (other than greenhouse gas emissions) of generic construction types and space heating energy usage. (Operating energy uses other than space heating depend on user behaviour and not on construction-related aspects and are, therefore, not included.)

Establishing the knowledge base of the model

Generic construction types based on the elements of a house

Architect-designed houses represent only a small percentage of houses in New Zealand, with the majority of new houses constructed being supplied by full-service housing companies (Vale et al. 2000). They offer three types of service:

1. Contract built houses – houses constructed based on standard plans, which are customised to the buyer's needs;
2. Design and build houses – unique houses designed to suit the requirements of individual buyers; and
3. Spec homes – a complete house built to sell.

Since the majority of new houses constructed in the lower-cost range of the market in the Auckland region are in the category of contract-built houses, a study was undertaken to identify the most common specifications and construction types used.

A sample of trade brochures, specifications, product technical information for specialised building systems/materials, BRANZ (Building Research Association of

New Zealand) appraisal certificates, and schedules of various plan/floor area/price options provided by construction firms, was studied based on the assumption that:

- study of information provided by the builders would give a complete picture of the general practice in the house building industry; and
- builders' brochures represent the most frequently used floor plans and designs.

Information extracted from the brochures was used as the means of identifying the construction methods used. Although the survey does not represent the actual market share since it is a snapshot of the industry, it does represent what is on offer to the house-buyer at a particular time and thus has a market focus rather than a manufacturers' focus. The results of this study are discussed next.

Common specifications used for building components
Foundation
The type of foundation used depends to a great extent on the site selected, with soil-bearing capacity, earthquake and wind loads being the governing factors (Figure 6.1). The most commonly used foundation types according to the study are:

- Tanalised timber piles (21%)
- house piles (timber/concrete) (31%)
- reinforced concrete footings (30%)
- Total (82%).

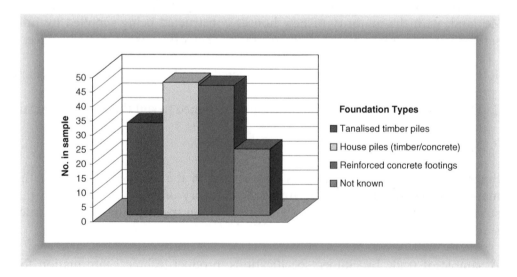

Fig. 6.1   Common foundation types used in New Zealand houses.

Pile foundations are generally referred to as house piles without the indication of whether they are in timber or concrete. Therefore, the category of house piles (31%) could also include Tanalised timber piles.

Floor
Ground floor construction is mainly of two types:

1. timber-framed suspended particleboard floor (65%);
2. concrete slab-on-gound floor with reinforced mesh (35%)

Timber-framed suspended floors are either:

- enclosed with a continuous perimeter foundation wall which excludes wind from the subfloor space; or
- exposed, allowing air circulation.

According to the minimum thermal insulation requirements for residential buildings, all houses need to satisfy certain performance standards (NZS 4218:1996). Although suspended floors would require some form of insulation to comply with this requirement, floor insulation is not a definite item in the builders' specifications studied. Therefore, it is not clear whether all the constructions comply with the standard.

Two types of floor insulation are currently being used (Figure 6.2):

1. Double-sided aluminium perforated foil insulation draped over the timber floor frame, which constitutes 58% of the sample;
2. 25 mm thick polystyrene perimeter insulation for concrete slab floors, which constitutes 8% of the sample.

In order to obtain the minimum floor R-value ($1.3 \, m^2C^0/W$) stipulated by the Code (NZS 4218:1996), foil insulation has to be laid with a sag of 100 mm below the floor. According to a previous field investigation carried out by BRANZ, it was found that 1 in 5 houses had foil insulation pulled too taut, leading to poorer R-values than expected (Isaacs and Trethowen 1985). Usually, floor covering is not included in the specifications, this being left to the client to choose at an extra cost. Most commonly used floor coverings are carpet and vinyl flooring for timber-framed floors and ceramic tiles for concrete floors.

External walls
According to published data (Page 1999: p. 9), the predominant specification for house wall construction is a light timber frame, with a market share of 94.5%. Although using steel framing in place of timber is advantageous in terms of

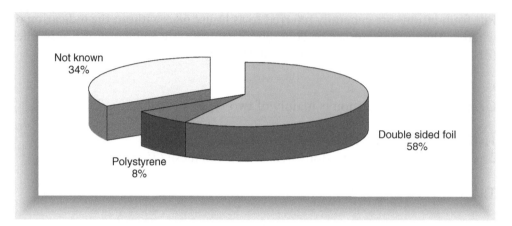

Fig. 6.2   Use of insulation in floor construction of New Zealand houses.

straightness, ease in carrying around due to light weight, immunity from moisture problems, etc., steel frames are poor in thermal performance and high in embodied energy content. Currently, light steel framing has only 1.5% of the market share, mostly in Auckland and Bay of Plenty, owing to slightly higher initial cost (Shelton 1999: p. 28). The present study also confirmed the above, though with a different range of percentages.

Wall construction types are (Figure 6.3):

- kiln-dried light timber frame (60%);
- double tongue and grooved (T&G) laminated timber (29%);
- precast insulated concrete wall panels (7%);
- rustproof steel frame (4%).

Three types of wall insulation materials are used:

1. glass fibre batts (58%)
2. polystyrene (7%)
3. double-sided aluminium foil (7%)
4. Total (72%).

Generally, glass fibre batts are placed between the studs of the timber wall frame. Internal wall lining is plasterboard with the external cladding varying as follows (Figure 6.4).

- fibre cement weather board on building paper (47%);
- brick veneer (13%);

- timber cladding (9%);
- fibre cement (F/C) backing board with a finishing coat (2%).

Therefore, 71% of all houses use some form of external cladding on a wall frame (timber or steel) or as an additional layer for better performance in the tongue and

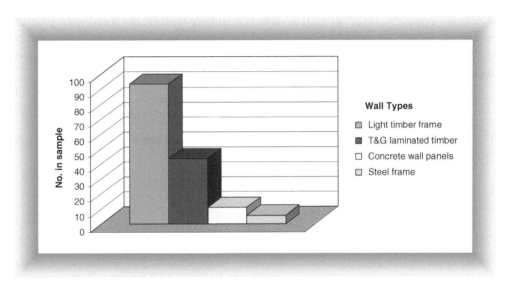

Fig. 6.3   Common wall types used for external walls of New Zealand houses.

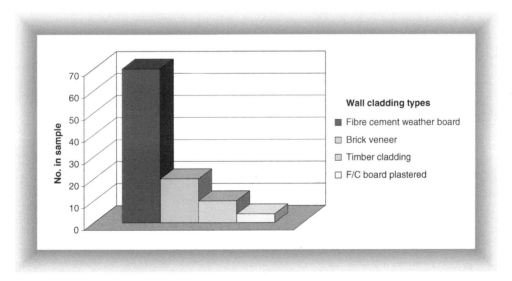

Fig. 6.4   External wall claddings used in New Zealand houses.

grooved interlocking timber wall system, with double-sided foil insulation and building paper. Precast concrete panels come with polystyrene insulation[1] and are finished with a solid plaster coat.

Internal walls

The most common internal wall system is a timber wall frame lined with plaster-board on either side (65%) finished with wallpaper or paint. In the system that uses concrete panels for external walls, internal walls are factory-built, wood-based panels.

Roof

For the pitched roof of a New Zealand house, three types of framing are used (Figure 6.5):

1. timber truss (53%)
2. timber rafters and beams (36%)
3. steel truss (11%).

As a result of the prefabricated framing systems in use, the timber truss is the most commonly used roof framing method. A rafter and beam system is mostly used for a timber ceiling with exposed roof beams, where interlocking wood construction forms the wall system.

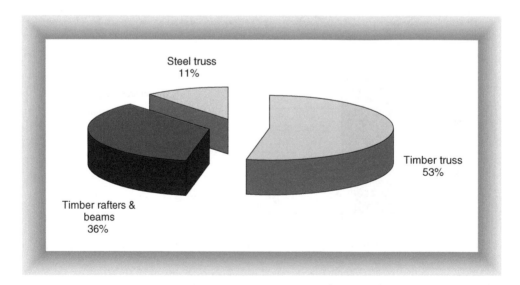

Fig. 6.5   Roof framing methods used in New Zealand houses.

Metal cladding is the prevalent roof covering (87%), though concrete roof tiles (12%) are closely associated with high-mass houses, i.e. concrete panel houses or timber frame with brick cladding (mass here refers to construction mass rather than to thermal mass). A flat ceiling with plasterboard lining (56%) in which bulk insulation is incorporated in the ceiling space to reduce heat loss from inside the house is the most commonly used internal ceiling finish. Wood fibre ceiling tiles (30%) are often used with exposed rafters, but not usually in houses with inter-locking wood construction.

The ceiling/roof area being the main route for unwanted heat loss means that the Building Code requires higher R-values ($1.5–3.0\,\mathrm{m^2C^0/W}$) in roof construction than in wall or floor construction. Fibreglass is used for 86% of roof insulation.

Joinery

Aluminium is predominantly used for external joinery and 99% of the houses in the sample studied use aluminium window frames with extensive areas of single glazing. The front entrance door, in addition to keeping out the rain and intruders, is used to welcome the guests and is often given some form of individuality. Commonly used front doors are as follows.

- glazed aluminium (52%)
- timber panel doors with natural/paint finish (43%)
- steel (5%).

According to the study, metal gutters are used in most cases (47%), closely followed by PVC (36%). In addition, 59% of the houses incorporated some form of decking in their designs, and 68% of the specifications included a hot water cylinder although only 5% of cylinders came with 'grade A' thermal insulation. According to the New Zealand Building Code, hot water cylinders must have a minimum of 25 mm of insulation and a thermal conductivity of less than 24 W/m°C when tested according to ASTM[2] C518-76 (Building Industry Authority 2000).

The results of the study are summarised in Appendix A.

The most common specification for a New Zealand house based on the above can be summarised as follows.

- particleboard floor on timber joists on Tanalised timber pile foundation, double-sided foil draped over floor frame as insulation;
- timber-framed walls with 94 mm of glass fibre insulation[3] within the framework. Internal lining is plasterboard with paint or wallpaper finish. External cladding is fibre cement or timber weatherboarding or brick veneer.

- pitched timber truss roof with corrugated metal cladding, flat ceiling lined with plasterboard, roof–ceiling space insulated with glass fibre.
- aluminium framed windows with single glazing.

Common construction types used in the New Zealand house

The most common specification used for NZ houses, therefore, is a raised light-weight timber-framed construction. There are several reasons for this, including the historical use of timber for residential constructions, the Earthquake Code requirements, which specify more complex engineering designs for masonry constructions, and the sloping terrain of most New Zealand sites, which is not conducive to the use of a concrete slab foundation. The specifications adopted for the most common construction used in the model hereinafter referred to as 'light construction' are as follows.

- particleboard floor on raised softwood framing, double-sided foil draped over floor frame as insulation;
- softwood framed walls with 94 mm of glass fibre insulation within the frame-work, plasterboard internal lining with paint finish, fibre cement external cladding;
- pitched soft wood truss roof with corrugated metal cladding, flat ceiling lined with plasterboard, roof–ceiling space insulated with 75 mm glass fibre;
- aluminium-framed windows with single clear glazing.

In the most commonly used masonry construction system in New Zealand, brick veneer is located on the exterior although, ideally, brick or other masonry should be inside the house to be profitable in terms of both thermal performance and the subsequent life cycle energy. Depending on the terrain, the floor construction of this masonry house varies from elevated timber frame to concrete slab on the ground. The specifications adopted for the 'high mass' version of the timber-framed house, hereinafter referred to as the 'heavy construction', are as follows.

- 150 mm thick concrete floor slab with 25 mm thick expanded polystyrene perimeter insulation to a depth of 500 mm (the thermal mass);
- softwood-framed walls with external brick veneer, plasterboard internal lining with paint finish, 94 mm of glass fibre insulation within the framework;
- pitched softwood truss roof with concrete tile covering, flat ceiling lined with plasterboard, roof–ceiling space insulated with 75 mm glass fibre;
- aluminium framed windows with single clear glazing.

In addition to the above, another three construction types used for house construction in New Zealand were identified and are as follows.

1. concrete slab on ground with a timber-framed envelope insulated to the same standard as the light construction above with a metal clad roof;
2. concrete slab on ground with 'mud brick' (i.e. soil–cement) walls and a timber-framed tiled roof (a construction which is not offered by mass-market house-builders but that is popular with some people who have houses constructed for them by individual building firms); and
3. solid interlocking timber construction with an insulated metal-clad roof.

In the common lightweight construction with extensive areas of single glazing used in New Zealand, condensation on the windows is a common problem during winter. However, owing to the high number of sunshine hours even during winter, condensation dries out on a daily basis when the sun is out (Table 6.1). These houses are also prone to overheating in summer, especially if the house is closed and unoccupied during the day.

In order to investigate the implications of the use of higher levels of insulation in New Zealand houses, a highly insulated (or superinsulated) hypothetical construction was also added to the model. This highly insulated construction doubled the insulation in the common light construction to achieve an R-value of $4.4\,\mathrm{m^2C^0/W}$ all around with double glazing for windows. Specifications adopted for the highly insulated construction, hereinafter referred to as 'superinsulated construction', are as follows:

- particleboard floor on raised softwood framing, with 200 mm of glass fibre on a plywood layer as insulation;
- softwood framed walls with 200 mm of glass fibre insulation within the framework, plasterboard internal lining with paint finish, fibre cement external cladding;
- pitched softwood truss roof with corrugated metal cladding, flat ceiling lined with plasterboard, roof–ceiling space insulated with 200 mm glass fibre;
- aluminium framed windows with double clear glazing.

The model is based on these generic construction types currently available in the market so that the model is in a format that building designers are familiar with. This database however, could be added to at any time to incorporate any future types of construction and could also be adapted easily for situations in other countries.

Embodied energy of NZ building materials, domestic appliances and furniture

The data on embodied energy of building materials used for the model are the most recent figures published for New Zealand building materials, with the

Table 6.1 Comparison of annual sunshine hours for main centres in New Zealand and the UK (based on 1970–2000 averages)

Location	Jan	Feb	Mar	Apr	May	Jun	Jul	Aug	Sept	Oct	Nov	Dec
NEW ZEALAND												
Auckland	229	201	180	162	142	111	140	144	149	181	187	225
Wellington	246	209	191	155	128	98	117	136	156	193	210	226
Christchurch	230	196	183	161	142	119	124	148	165	198	215	221
Dunedin	178	153	140	121	100	86	101	114	129	147	161	169
Queenstown	228	206	189	141	91	75	87	121	158	189	208	227
Invercargill	180	165	136	110	80	76	91	119	134	155	176	186
UK	43	64	96	140	184	168	173	165	123	91	57	36

Based on: http://www.metoffice.gov.uk/climate/uk/averages/19712000/areal/uk.html, accessed 21 January 2007; http://www.niwascience.co.nz/edu/resources/climate/sunshine/, accessed 20 January 2007 widest coverage (Alcorn and Wood 1998). Although a more recent set of values is available (Alcorn 2003), it is limited in its coverage and therefore was not used. Embodied energy figures used for the model are as shown in Appendix B. This shows how the values have changed over the years.

Embodied energy of domestic appliances and furniture

The embodied energy of domestic appliances and furniture was not included in the earlier energy studies carried out in New Zealand, and therefore the embodied energy of appliances and furniture used in New Zealand had to be established for the model. Fay (1999) used 10 and 8 MJ/A$ as the embodied energy intensity of domestic appliances and furniture respectively, based on Treloar (1996). However, a simple calculation of the embodied energy of a wooden coffee table on sale in New Zealand, based on the mass of the material, revealed that it comprised a total embodied energy of 61 MJ leading to an energy intensity of 0.2 MJ/NZ$ (note: The NZ dollar currently has a value roughly 80–90% that of an Australian dollar).

On inquiry, Treloar (2000)[4] conveyed that his figure was based on 1992–1993 input–output tables, and consists of primary energy as follows.

Direct energy	15%
Iron and steel	10%
Basic non-ferrous metal and products	8%
Plywood, veneers and fabricated wood	8%
Road transport	3%
Others	53%

This confirms that Treloar's figure is for an average item of furniture, examples of which often consist of a combination of materials, leading to variations due to the differences in the manufacturing processes. Further, the value of 0.2 MJ/NZ$ was calculated based on a table made of solid timber, which has a relatively low embodied energy (2,020 MJ/m^3) compared with high-energy materials such as iron and steel (97,890 MJ/m^3), and does not include transport and many other third-party energy components included in the Australian figures under the 'others' category, which in this case constitutes 53% of the total. Solid timber does not, indeed, appear in Treloar's table of values. The difference between Treloar's figures and the value calculated for the wooden coffee table emphasises the point that the materials from which furniture is made can make a significant difference to its embodied energy.

For a similar study investigating the life cycle performance of Australian houses, Fay et al. (2000, p. 36) used 8 MJ/A$ as the embodied energy of both the appliances and furniture while Fay (1999: p. 372) used 10 MJ/A\$ for appliances and 8 MJ/A\$ for furniture. In order to try to establish appropriate values for the embodied energy of the appliances and furniture used in an average house in New Zealand, the embodied energy coefficients of some common building materials in New Zealand and Australia were compared, as shown in Table 6.2.

The figures shown in Table 6.2 indicate that the embodied energy intensity for Australia is around 8 times greater for steel and copper, while also being about 6 times greater for timber, in comparison with New Zealand. Although the wide variation in the system boundaries used for energy analysis in New Zealand and Australia, such as inclusion of third-order items (e.g. banking and insurance) in the figures for Australia, could lead to such a situation, the disparity between the two economies also contributes. In Australia, with an energy industry that is heavily dependent on coal, 3.4 MJ of primary energy is used to produce 1 MJ of delivered energy as electricity (Fay et al. 2000: p. 33). However, according to Baines and Peet (1995), quoted by Alcorn (2003: p. 9), only 1.53 MJ of primary energy is used to generate and distribute 1 MJ of electricity (delivered energy) in the New Zealand energy industry. Owing to the lack of information representative of the New Zealand manufacturing and distribution system, it could be concluded (noting that the figures for copper and steel in Australia are 8 times greater than those for New Zealand and that the value of the NZ dollar is 80% of the value of an Australian dollar) that the embodied energy of New Zealand-made domestic appliances is 1.0 MJ/NZ\$, i.e. one eighth of the corresponding figure for Australia. On the same basis, noting that the figure for timber for Australia is 6 times that for New Zealand and taking into account the relative values of the NZ and Australian dollar, the embodied energy of New Zealand-made domestic furniture is 1.06 MJ/NZ\$, i.e. one sixth of the corresponding figure for Australia. These figures were consequently used in the model to represent the embodied energy of appliances and furniture.

Lamps

The embodied energy of lamps (i.e. light bulbs) was estimated based on a price of 1.20 NZ\$ for incandescent lamps and 7.00 NZ\$ for comparable compact fluorescent lamps, and at an embodied energy intensity of 1.0 MJ/NZ\$.

Replacement cycles for building elements/materials, appliances and furniture for a New Zealand house

The number of times the components have to be replaced during the lifetime of the building is given by the formula:

$$\text{(Useful life of the building/useful life of the component)} - 1$$

Table 6.2 Comparison of embodied energy intensity of building materials in New Zealand and Australia

Material		New Zealand Alcorn & Wood (1998)	Australia Fay (1999)	Ratio Australia/NZ
Aluminium	virgin	191 MJ/kg	264.2 GJ/t	1.4
Aluminium	foil	55.08 GJ/m³	0.27 GJ/m²	1.0
Bitumen		45.42 GJ/m³	10.83 GJ/m³	0.24
Cement		9.0 MJ/kg	13.07 GJ/t	1.52
	fibre cement board 7.5 mm/6.0 mm	116.6 MJ/m²	0.32 GJ/m²	2.7
Ceramic	brick	5170 MJ/m³	0.88 GJ/m²	1.5
Concrete, ready mix	30 MPa	3180 MJ/m³	5.85 GJ/m³	1.8
Copper		70.6 MJ/kg	607 GJ/t	8.6
Glass	Float 6 mm/4 mm	240 MJ/m²	0.39 GJ/m²	1.6
	Toughened 6 mm	396 MJ/m²	0.91 GJ/m²	2.3
Insulation	Glass fibre batts	970 MJ/m³	0.25 GJ/m²	2.6
Paint		6.5 MJ/m²	0.02 GJ/m²	3.0
Plastic		103 MJ/kg	308.39 GJ/t	3.0
Plasterboard	9.5 mm	55.95 MJ/m²	0.14 GJ/m²	2.5
Sand		230 MJ/m³	0.33 GJ/m³	1.4
Steel	Stainless steel	50.4 MJ/kg	377.4 GJ/t	7.5
Timber, softwood	kiln dried, dressed	2204 MJ/m³	11.69 GJ/m³	5.3
	MDF	8330 MJ/m³	8.52 GJ/m³	1.02
Vinyl flooring		105.9 GJ/m³	0.32 GJ/m²	2.0

Based on: Alcorn, A. and Woods, P. (1998); Fay, M. R. (1999)

Replacement cycles for building elements and materials
The New Zealand building code (NZBC) requires residential buildings to have a serviceable life of 50 years (some historical buildings in New Zealand, such as Kemp House [1821–1822], have lasted for much longer). The minimum lifetime requirement for building elements in the NZBC is based on their function and accessibility for repairs. The structural elements and those items that cannot be accessed and detected during general maintenance are required to have a minimum life of 50 years. The required minimum life for elements that are easy to access and replace is 5 years, while it is 15 years for all other elements.

In order to attempt to determine the actual useful life of building materials and elements used in an NZ house, the previous works by Jaques (1996), Adalberth (1997), Fay (1999) and Johnstone (2001) were studied. Jaques, in his review of the embodied energy of a typical NZ house, considered a lifetime of 50 years while Johnstone used 90 years in his life cycle analysis of an NZ house. Adalberth took 50 years to be the useful life of a Swedish house while Fay used the figure of 100 years for an Australian house, in their life cycle analyses. Rawlinsons (1998), which publishes an annual price book for the Australasian building industry, has published the useful-life estimates for building materials and elements assuming a useful building life of 50 years. Since Rawlinsons' figures are intended to be used for tax depreciation purposes, they do not represent actual useful life.

Jaques (1996) assumed the following maintenance and replacement cycles for the main building elements based on expert opinion.

- Repaint roof once every 7.5 years.
- Repaint external walls once every 10 years.
- Replace wall paper and sheet vinyl flooring once every 15 years.
- Replace shower, taps and sink unit once every 20 years.
- Replace kitchen stove, hot water cylinder, aluminium window frames and spouting (guttering) once during 50 years.

The useful-life estimates used in the study of Fay et al. (2000) are: paint, 10 years; windows, 50 years; plumbing and electrical services, 25–75 years; appliances, 13–25 years; and roofing materials, 25–50 years. Table 6.3 gives a comparison of useful life estimates by Adalberth, Jaques, Fay, Rawlinsons and Johnstone.

The figures shown in Table 6.3 reveal that there is no commonly accepted useful life for various materials and elements used and estimates are purely based on the

Table 6.3 Comparison of useful life of building materials and elements using estimates from previous research

Building component	Materials	Adalberth (1997)	Jaques (1996)	Rawlinsons (1998)	Fay (1999)	Johnstone (2001)
Substructure	Timber piles, concrete slab	50	–	–	>100	40
Floor	Floor framing, joists, flooring	50	–	–	>100	–
Walls	Timber studs and wall framing, plasterboard	50	–	–	>100	40
	Insulation, skirting, brickwork, mortar, cavity ties, flashings	50	–	–	>100	–
	Wooden panelling	30	–	–	–	–
	Fibre cement weatherboard	–	–	–	–	50
	External rendering	–	–	–	60	50
Roof	Timber/steel roof frame	50	–	–	>100	–
	Plasterboard ceiling lining & battens	–	–	20	>100	–
	concrete tiles & battens	30	–	–	>100	–
	Steel roofing sheets, battens, insulation	–	–	–	40	50
	Gutters and downpipes	30	30	–	20	25
Electrical work	Wiring, switchboard & power outlets	50	–	–	50	40
Joinery	Window frames, glazing	30 (timber)	30 (aluminium)	–	60 (aluminium)	40 (aluminium)
	External doors, frames, architraves	30	–	–	60	20
	Internal doors, frames	30	–	–	60	–
	Door and window furniture	–	–	–	60	60

(Continued)

Table 6.3 (Continued)

Building component	Materials	Adalberth (1997)	Jaques (1996)	Rawlinsons (1998)	Fay (1999)	Johnstone (2001)
Plumbing	Hot water service	16	30	12.5	16	16
	Sanitary fittings – basins, sinks, baths, shower trays, tapware	–	20	8	30	20
	Copper, PVC & UPVC pipes	50	–	25	50	–
Finishes	Replace vinyl flooring	17	15	10	30	10
	Replace parquet flooring	50	–	15.5	–	–
	Replace ceramic floor tiles	–	–	–	30	–
	Replace wool carpets	–	–	5	12	10
	Replace wall papering	10	15	–	–	8
	Repaint cladding	–	10	–	8	8
	Repaint doors, trim and ceiling	10	–	–	8	8
	Replace curtains	–	–	8	–	–
	Repaint roofing	10	7.5	–	–	7
	Kitchen upgrade	30	–	–	30	25

Sources: Adalberth, K. (1997); Jaques, R. (1996); Rawlinsons (1998); Fay, M. R. (1999) and Johnstone, I. M. (2001)

experience of individuals. Owing to advancements in technology, the aluminium framing for doors and windows commonly used in New Zealand today is powder coated. Based on anecdotal evidence the durability of windows has increased from around 50 years to approximately 60 years and Jaques' estimate seems very low. However, based on expert opinion, Jaques' assumptions as regards the durability of the hot water cylinder and kitchen stove seem to be very high. The generally accepted lifetime for a hot water cylinder seems to be around 15 years according to four out of the five studies considered above. Fay's estimate of the durability of vinyl flooring is also very high. Adalberth's estimates are for the colder climate in Sweden, whereas those of Fay are for the hotter and drier climate in Australia, and this could also affect the useful life.

In addition to the above replacement cycles, Johnstone (2001) calculated the annual maintenance of a typical New Zealand house based on the maintenance records of a sample of 25 houses maintained by Housing New Zealand, which is the national provider of social rental housing. His annual maintenance requirements included the following.

- repairs to hot water cylinder;
- repairs to electrical outlets, lighting and meter board;
- replacement of taps and washers;
- clearing of blockages in drainage systems;
- repair and replacement of flashing;
- repairs to fittings, cupboards and shelving.

Johnstone (2001: p. 34) estimated the cost of the activities listed above to amount to 0.14% of the cost of constructing a new house. The sample used for this purpose was small, and, as it consisted of government housing, the maintenance schedules and the way the tenants use such houses may not be representative of the general picture in New Zealand. (A survey of owner-occupied New Zealand housing in 2005 [Clark et al. 2005] found that the level of house maintenance is generally insufficient to sustain satisfactory conditions.) Further, items such as repairs to a hot water cylinder, blockages in a drainage system, repairs to flashings and fitting cupboards, do not normally occur annually in a normal house. Some of these repairs identified by Johnstone may well be the result of occupant behaviour rather than the failure of the material/element. Models cannot predict aberrant occupant behaviour. In a life cycle analysis, such influences have to be omitted.

The replacement cycles of building elements and components used for this model are as shown in Table 6.4. While this information was drawn from several

Table 6.4   Replacement cycles for building components and elements

Building component	Materials	Replacement cycle		
		Best	High	Low
Substructure	Timber piles	>100	50	>100
	Concrete slab	>100	>100	>100
Floor	Floor framing, joists, flooring	>100	50	>100
Walls	Timber studs and wall framing, plaster board	>100	50	>100
	Insulation, skirting, brick work, mortar, cavity ties, flashings	>100	50	>100
	Fibre cement weatherboard	50	40	60
	Wooden panelling	30	20	40
	External rendering	60	50	75
Roof	Timber/steel roof frame	>100	50	>100
	Plasterboard ceiling lining & battens	>100	20	>100
	Concrete tiles & battens	>100	30	>100
	Steel roofing sheets, battens, insulation	40	30	50
	Gutters and downpipes	20	15	30
Electrical work	Wiring, switchboard & power outlets	50	40	60
Joinery	Aluminium window frames, glazing	60	30	65
	External doors, frames, architraves	60	20	65
	Internal doors, frames	60	30	65
	Door and window furniture	60	40	65
Plumbing	Hot water service	16	12.5	30
	Sanitary fittings – basins, sinks, baths, shower trays, tapware	30	20	40
	Copper, PVC & UPVC pipes	50	25	60
	Towel rail, toilet paper holder	20	15	30
Finishes	Replace vinyl flooring	17	10	30
	Replace parquet flooring	50	15.5	60
	Replace ceramic floor tiles	30	20	40
	Replace wool carpets	12	5	15
	Replace wall papering	10	8	15
	Repaint cladding	8	6	10

Table 6.4 (Continued)

Building component	Materials	Replacement cycle		
		Best	**High**	**Low**
	Repaint doors, trim and ceiling	8	6	10
	Replace curtains	8	6	10
	Repaint roofing	10	7.5	12
	Kitchen upgrade	30	25	40

Based on: Adalberth, K. (1997); Jaques, R. (1996); Rawlinsons (1998); Fay, R. (1999); Johnstone, I. M. (2001)

sources, where information is lacking intelligent judgement has been used. The 'best' estimate is for maintenance of average standard while 'high' replacement is for maintenance of an above average standard, which might be wasteful of materials, and 'low' is for a below-average standard, which might lead to degradation of the building and hence a shorter life.

Replacement cycles of domestic appliances and furniture

The estimates for the useful life of domestic appliances vary depending on the data source. While Adalberth (1997: p. 318) estimated this to be 12 years on average, Rawlinsons' (1998: p. 595) estimate is about 8 years. Since estimates by Rawlinsons are for tax depreciation purposes, they may not represent the actual useful life of appliances. Fay (1999: pp. 373–374) has estimated a useful life of 21 years for a refrigerator/freezer, 18 years for both the electric range/oven and the clothes dryer, 17 years for a refrigerator and 15, 14 and 13 years for the water heater, washing machine and dishwasher respectively. The replacement cycles for domestic appliances and their purchase value used for the model are as shown in Table 6.5.

The useful life of furniture was assumed as 25 years for the purposes of the model based on Fay (1999). For lamps a useful life of 1,000 hours for incandescent lamps and 5,000 hours for compact fluorescent lamps was assumed, and the replacement needs were calculated based on the number of hours of use in each living area as published by Wright and Baines (1986).

It should also be noted that many appliances and items of furniture are often discarded not because they have failed, but because they are no longer considered to be in style, or are no longer technologically up to date. This trend is quite likely to increase.

Table 6.5 Domestic appliances used in New Zealand houses, their useful life and purchase value (2006)

Appliance	Useful life	Value (NZ$)
Electric range and oven	15	800.00
Microwave oven	12	250.00
Refrigerator/freezer	17	900.00
Deep freezer	17	700.00
Toaster	8	40.00
Electric kettle	8	90.00
Washing machine	14	800.00
Clothes dryer	18	400.00
Colour television	10	600.00
DVD/video player	10	250.00
Iron	8	90.00
Vacuum cleaner	8	250.00
Home computer	5	1,900.00

Based on: Fay, R. (1999) Trade literature – Betta Electrical, October 2006

Installed prices of building materials/components and price of energy used in a New Zealand house

Average installed prices for a wide range of building materials and constructions on a unit basis for six main centres are published by the New Zealand Building Economist (Wilson 2006). The rates used for this analysis are based on the information published in August 2006, for Auckland. These prices do not include GST (Goods and Services Tax). Therefore GST at the current rate of 12.5% has been added to these figures.

The price of appliances is as shown in Table 6.5. The furniture used in the house and its purchase value were assumed as shown in Table 6.6.

The cost of electricity used for operating energy was calculated based on the standard user charges by Mercury Energy (February 2007). A line charge (which covers the cost of being connected) of 72.26 cents per day and a unit charge of 15.95 cents/kWh were used. The above charges do not include GST, and hence it has been added (at 12.5%) to the final cost.

Table 6.6   Furniture for the NZ house and purchase value (2006)

Item	Value (NZ$)
Three-piece lounge suite	2,000.00
Coffee table	300.00
2no. book cases	300.00 each
TV stand	350.00
Seven-piece dining suite	1,200.00
Double bed	1,400.00
2no. single beds	1,000.00 each
2no. chests of drawers	500.00 each
Writing desk & chair	500.00

Source: Trade literature – Furniture City, October 2006

Operating energy requirements of space conditioning, appliances, lighting and hot water system

Space conditioning

Space conditioning energy use has to be calculated separately using a thermal simulation programme and transferred to the life cycle model. (Since cooling is not currently a common practice in an NZ house, space conditioning energy use is for space heating during winter.) The actual energy use is, however, also dependent on both the heater type and the fuel type used. These were assumed based on the information published by Camilleri (2000b) and are as shown in Table 6.7. Assumptions regarding heating are discussed in Chapter 7.

Appliances

The total electrical energy used for domestic appliances in an average Auckland house according to HEEP (Household Energy End-use Project, a 10-year research programme to discover how energy is being used in New Zealand homes) monitoring data (Isaacs 2004) is 2,686 kWh per annum. However, this does not reveal appliance usage patterns. Wright and Baines (1986) predicted energy usage patterns in average NZ houses in the year 2000. Domestic appliance usage patterns and respective electricity consumption figures used for the model are based on a combination of the findings of the HEEP study and Wright and Baines (1986). These are shown in Table 6.8.

The appliances listed in the Table 6.8 use 2,627 kWh per annum and the balance of 59 kWh per annum is assumed to be used by minor equipment such as clock–radios,

Table 6.7   Efficiency and CO_2 emissions due to use of various heating appliances

Fuel	Heater type	Fuel emission factor (kg CO_2 eq./ kWh)	Efficiency	Kg CO_2 equivalent/ kWh heating output
Electricity	Air conditioner	0.64	1.90	0.34
Electricity	Ducted heat pump	0.64	1.68	0.38
Electricity	Resistance	0.64	1.00	0.64
Electricity	Floor	0.64	0.90	0.71
Electricity	Night store	0.64	0.80	0.80
Electricity	Ceiling	0.64	0.60	1.07
Natural gas	Unflued	0.19	0.81	0.23
Natural gas	Flued	0.19	0.80	0.24
Natural gas	Central heating	0.19	0.66	0.29
LPG	Unflued	0.22	0.81	0.23
LPG	Flued	0.22	0.8	0.24
LPG	Central heating	0.22	0.66	0.29
Diesel	Central heating	0.25	0.42	0.59
Coal	High-eff. double burner	0.36	0.8	0.44
Coal	Basic double burner	0.36	0.65	0.55
Coal	Pot belly	0.36	0.35	1.01
Coal	Free standing metal fire	0.36	0.25	1.42
Coal	Open fire	0.36	0.15	2.37

Source: Camilleri, M. J. (2000) *A Draft Climate Change Sustainability Index for Houses.* BRANZ Study Report 95, Building Research Association of New Zealand, p. 15

alarms and mobile-phone chargers. The total electricity consumption for appliances is calculated by the model based on the number of appliances used in the house.

Lighting

According to the HEEP study, average electricity use for lighting in an Auckland house is 1,185 kWh per annum. Domestic lighting usage patterns built into the model based on the figures predicted by Wright and Baines (1986) are as shown in Table 6.9.

Table 6.8   Domestic appliance usage pattern in an NZ house

Appliance	Wattage	Usage	kWh/year
Refrigerator/freezer	–	–	570
Deep freezer	–	–	436
Toaster	850	36 min/week	27
Electric kettle	2,200	80 min/week	153
Washing machine	–	–	47
Clothes dryer	–	–	454
Colour television	154	6.25 hours/day	351
Video recorder	23	2 hours/day	17
Iron	2,400	1.5 hours/week	187
Vacuum cleaner	1,800	1.5 hours/week	140
Computer	100	2 hours/day	73
Microwave	1,100	3 hours/week	172

Table 6.9   Electrical lighting usage in an NZ house (using incandescent lamps)

Location	Wattage	Hours/day
Living/dining		4.0
– large	3×60	
– small	2×75	
Kitchen		6.0
– large	2×75	
– small	1×100	
Master bedroom	1×100	0.5
Second bedroom	1×75	0.5
Hallway	1×100	2.5
Bathroom	1×75	0.75
External access	1×100	5.0

Water heating energy

According to the HEEP study (Stoecklein et al. 2002), electrical energy use for water heating in an average NZ house is 4,000 kWh per annum. This is used in the model to represent the operating energy requirements of water heating for up to three occupants. For larger houses, water heating energy use is calculated by the model based on the number of occupants.

Cooking

Average electricity use in an Auckland house for the kitchen stove (including oven and hob) as measured by the HEEP study (Isaacs, 2004: p. 7) is 474 kWh per annum. This was used in the model to represent electrical energy use for cooking.

Greenhouse gas emissions due to NZ building materials, appliances and furniture

Unlike other building-related environmental impacts, greenhouse gases are location-specific (as discussed in the Conclusions section of Chapter 5), and greenhouse gas emissions at various stages of the NZ house that could lead to global warming need to be considered. Greenhouse gas emissions due to chemical reactions in the production of the main building materials (cement, steel, aluminium and lime) used in NZ houses are shown in Table 5.3. In addition to the above, carbon dioxide is also emitted due to the process energy used in the manufacture of these and other materials.

The CO_2 equivalent greenhouse gas emission factors for building materials that have been published (Alcorn 2003) are limited in their coverage. For accurate calculation of process energy based CO_2 emissions, data on energy mix used in the production process are necessary. This information, however, is not readily available. Mithraratne (2001) derived CO_2 equivalent greenhouse gas emission factors for NZ building materials based on embodied energy figures published in 1996 and energy sector-based CO_2 emissions for the year 1999. These values suffer from the lack of information on the actual energy mix, and carbon locked in timber products not being considered. Owing to variations in the base dates, highly variable energy (electricity)-related emissions in New Zealand and the analysis methods used, the two data sets (Alcorn 2003 and Mithraratne 2001) are not comparable.

Embodied energy values for some of the building products have been updated in the 1998 data set used for the present analysis. Hence, CO_2 emissions for building materials published in Mithraratne 2001 were updated using the 1999 energy-related CO_2 emissions factor (0.0572 kg/MJ). Carbon dioxide emissions additional to process energy emissions are adjusted as follows.

Based on Wainwright and Wood (1981: p. 78), $1 m^3$ of cement mortar (1:3 mix), and concrete (1:2:4 mix) contain $0.338 m^3$ and $0.214 m^3$ of cement respectively. Based on an average weight of cement of $1,420 kg/m^3$, a cubic metre of cement mortar and concrete contain 480 and 304 kg of cement, respectively. (Honey and Buchanan used 320 kg of cement per $1 m^3$ of concrete based on a previous study.)

According to information from James Hardie Building Products (2001)[5], the cement content of fibre cement boards is about 30% by weight. This has been used to adjust the CO_2 emissions due to fibre cement boards. Carbon dioxide emissions due to ready mixed concrete were calculated based on information received from Ready Mixed Concrete (2001)[6] in Auckland. Ready mixed concrete grades 17.5, 30 and 40 Mpa contain 200–220, 280–310 and 330–350 kg/m^3 of cement, respectively. Carbon dioxide emissions have been calculated based on $210 kg/m^3$ for 17.5 Mpa, $295 kg/m^3$ for 30 Mpa and $340 kg/m^3$ for 40 Mpa concrete.

The cement content of cement-stabilised earth has been calculated based on the following quotation from Williams-Ellis et al. (1947: p. 68).

> *With good Pisé soils, that is, soils of a predominantly sandy type and containing only about 25% to 30% of fine material, between 4% to 7% of dry cement by weight is required in the mix to give a hardening effect to the soil. The more clayey the soil, the more normally unstable it will be, and consequently the more cement will be required to give a satisfactory stabilising and hardening effect. The clayey soils may be substantially hardened by the addition of between 6% and 10% of dry cement.*

While 5% cement by weight is used for sandy earth, 8% cement by weight is used for clayey earth. Eight per cent cement by weight has been used here to calculate cement content in soil cement.

Embodied energy of virgin aluminium used in New Zealand is 191 MJ/kg (Alcorn 1998: p. A1). According to Alcorn (2001)[7] this figure includes the energy used in Australia to extract bauxite, and to process bauxite into alumina and transport it across to New Zealand. Therefore, total CO_2 emissions attibutable to aluminium consist of process energy-based CO_2 (both in Australia and New Zealand), and chemical process emissions in Australia and New Zealand. According to the Ministry for the Environment (2000: p. Processes98, co2) chemical process emissions inside New Zealand are 1.704 kg CO_2/kg. In addtion to this, CO_2 is emitted in Australia during the process of smelting bauxite. According to Honey and Buchanan (1992: p. 29), the smelting of bauxite releases 130 kg of carbon per tonne of aluminium. Therefore, 0.477 kg CO_2 per kg aluminium ($0.130 \times 44/12 = 0.477$) is emitted in Australia. Although CO_2 emissions due to process energy use within Australia could be higher than in New Zealand (Honey and Buchanan 1992: p. 82), due to lack of data on energy use the New Zealand energy-based CO_2 figure (0.0572) was used to derive process energy-based CO_2 emissions. Therefore, total CO_2 emissions due to the manufacture of aluminium are:

$$191 \times 0.0572 + 1.704 + 0.477 = 13.1 kgCO_2/kg$$

In addition to CO_2, the manufacture of cement and steel emits limited quantities of carbon monoxide (CO) and oxides of nitrogen (NO_x). While the lifetime of NO_x is limited to several days, for CO it is still only several months. Therefore, the effect of the limited quantities of these two gases could be neglected. However, manufacture of aluminium emits perfluorocarbons[8] (CF_4 – 0.0083/317.6 kt/kt; and C_2F_6 – 0.0008/317.6 kt/kt) and SF_6 (0.0001/317.6 kt/kt), compounds with high global warming potentials (MfE[9] 2000: p. Processes 98, nonco2). Although the quantities are limited, this has been included in the total greenhouse gas emissions. Therefore, total CO_2 equivalent greenhouse gas emissions due to the manufacture of virgin aluminium are:

$$13.1062 + (0.0083/317.6) \times 6,500 + (0.0008/317.6)$$
$$\times 9,200 + (0.0001/317.6) \times 23,900 = 13.1062$$
$$+ 0.16965 + 0.023 + 0.00717 = 13.3 \ kgCO_2 \ equiv./kg$$

Wood takes from the air, and stores, 250 kg of carbon per cubic metre (Honey and Buchanan 1992: p. 30). Hence, the use of 1 m^3 of wood locks away 916.67 kg CO_2. This has been accounted for in the emissions associated with timber products. The CO_2 equivalent greenhouse gas emission factors for building materials updated and used for this model are given in Appendix C.

Greenhouse gas emissions due to appliances and furniture

Figures for total energy used in New Zealand and total CO_2 emissions have been published. New Zealand used 432,923 TJ of energy with 29,706 Gg of actual CO_2 emissions in 2004 (MfE 2006: p. 141). While this total energy usage includes all forms of fossil fuels (liquid, solid and gas), it does not include wood burned or CO_2 emissions resulting from this process. (Wood takes carbon from the air and stores it, releasing this as CO_2 when the wood rots or is burned. Therefore, CO_2 emissions due to the burning of wood could be omitted, as they are part of a continuous cycle.) The above figures lead to a CO_2 emission factor for the New Zealand energy sector of 0.0686 kg CO_2/MJ.

Although it is assumed here that all types of energy emit uniform amounts of CO_2, this is clearly not the case in practice. Electricity-related CO_2 emissions in

New Zealand due to the use of average (roughly 70% hydro) and thermal (natural gas and coal) electricity are 0.0278 (0.1 kg CO_2/kWh) and 0.1778 (0.64 kg CO_2/kWh) kg CO_2/MJ respectively (Camilleri 2000a: p. 58). The CO_2 equivalent greenhouse gas emission factor for energy used in the model is 0.0686 kg CO_2/MJ based on the above calculation, which covers all direct energy uses, not just electricity. This is because the energy mix for the manufacture of appliances and furniture is not generally known. Therefore, CO_2 equivalent greenhouse gas emissions for appliances and furniture are calculated based on their embodied energy and the above CO_2 factor.

Carbon dioxide emissions due to electricity used for lighting, cooking and domestic appliances are calculated by the model using the total electricity use and the conversion factor of 0.64 kg CO_2/kWh of electricity. (Although some of this electricity use, such as for the refrigerator, would be at off-peak time, any additional demand for electricity in the evening and early morning when most appliances are used has to be met by thermal power generation [Camilleri 2000a: p. 55] and therefore the use of this higher CO_2 emission figure for electricity is justifiable.)

Environmental impacts other than greenhouse gases due to construction

Environmental impacts due to building construction other than greenhouse gas emissions are not location specific. Impacts such as those listed below are dependent on the production process and, therefore, global information could be used to analyse these environmental impacts attributable to NZ houses.

- recycling/disposal of building materials;
- water use for building material production and construction;
- pollution caused by hazardous waste;
- VOCs and other building-related health problems.

Very limited information exists on the environmental impact of New Zealand building materials.

Woolley et al. (1997) have published data on environmental impacts of building materials and techniques used in the UK. Owing to the difficulty of assigning comparable units, the above information, however, is insufficient to enable accurate quantification of environmental impacts other than greenhouse gas emissions. Hence, a rating system indicative of the impact of the generic construction types used in NZ houses was derived based on Woolley and colleagues' data; these are shown in Table 6.10.

Table 6.10   Environmental impact rating for generic construction types

Generic construction types	Rating
Foundation	
Timber piles on concrete footing	1
Concrete piles on concrete footing	2
Reinforced concrete continous footing	3
Floor construction	
Timber framed with aluminum foil insulation and particle board flooring (R = 1.33)	1
Timber framed with 200 mm of glass fibre insulation and 3 mm plywood and particle board flooring (R = 4.4)	2
Reinforced concrete slab (R = 1.62)	3
External wall construction	
Tongue & grooved solid timber	1
'Earth brick' wall	2
Timber-framed glass fibre insulated with fibre cement weather board cladding (R = 2.2)	3
Timber-framed 200 mm glass fibre insulated with fibre cement weather board cladding (R = 4.4)	4
Timber-framed glass fibre insulated with brick veneer (R = 2.1)	5
Roof construction	
Timber-framed concrete tiled roof glass fibre insulated with flat gypsum plaster board ceiling (R = 1.8)	1
Timber-framed metal-clad roof with glass fibre insulated flat gypsum board ceiling (R = 1.9)	2
Timber-framed metal clad roof with 200 mm glass fibre insulated flat gypsum board ceiling (R = 4.4)	3
Floor finishes	
Parquet flooring	1
Ceramic floor tiles	2
Wool carpets	3
Vinyl flooring	4
Wall finishes	
Wall papering	1
Wall painting	2

A score of 1 indicates the least impact among the types considered for a particular element, with higher ratings indicating higher impact. Although the listing in Table 6.10 embraces the generic constructions currently used in New Zealand, it is extendable to include any future additions. This would, in theory, allow designers to decide on the generic construction type to be used in NZ houses based on environmental impact, but the process involved would be ineffective because the list does not include the operating requirements of the house, which will make a significant contribution to the total impact. It might well make sense to increase the initial impact in order to decrease the overall life cycle impact.

This system of rating, coupled with the percentage composition of life cycle embodied energy of NZ houses, is used in the model to provide an indication of impact. The impact of the use of a certain generic construction type would be the assigned rating multiplied by the percentage of the item in the total embodied energy. Although this is not a quantitative assessment, with the limited information currently available this seems to be the best available assessment.

Environmental impacts other than greenhouse gases due to space heating

In rating the space heating energy requirement, the most common construction and its heating requirement are considered as the basis and assigned the poorest rating. As the purpose of rating is to promote better performance, the constructions using the current standard practice would only be able to achieve the lowest rating of 6. The environmental impact rating for space heating energy use used in the model is as shown in Table 6.11.

When space heating requirement is included, the environmental impact would be calculated using the composition of the life cycle energy rather than the composition of embodied energy. The impact would be the product of environmental impact rating times the percentage in the life cycle energy.

User interface of the model

The graphical user interface allows the user to communicate with the model by selection and input of data and consists of a series of forms to be filled in, based on the quantities of material required to make the house, which is in turn based on the relevant building elements (foundation, floor, walls, etc.). The space heating energy requirement has to be separately calculated and transferred.

Table 6.11   Environmental impact ratings for space heating energy use

Space heating requirement	Rating
More than 85% of the code requirement	6
Less than 85% of the code requirement but more than or equal to 65%	5
Less than 65% of the code requirement but more than or equal to 50%	4
Less than 50% of the code requirement but more than or equal to 35%	3
Less than 35% of the code requirement but more than or equal to 20%	2
Less than 20% of the code requirement but more than 0%	1
Zero space heating energy	0

Inference engine of the model

The inference engine contains the rules and formulae necessary to derive information from the knowledge base to provide appropriate responses to user input and selections.

The model with an earlier (1996 based) data set was used by the students of the School of Architecture at the University of Auckland for design evaluation in early 2000. In addition, the model has been validated using comparative energy studies of New Zealand residential buildings (Mithraratne 2001). The use of this model to evaluate life cycle performance of common NZ house types is discussed in the next chapter.

Notes

[1] It is not clear however, whether insulation is generally internal or external.
[2] American Society for Testing and Materials.
[3] Due to ease of construction, wall insulation is thicker than ceiling insulation.
[4] Personal communication.
[5] Personal communication.
[6] Personal communication.
[7] Personal communication.
[8] Total perfluorocarbon emissions due to production of aluminium (317.6 kt) in New Zealand during 1998–1999.
[9] Ministry for the Environment.

References

Adalberth, K. (1997) Energy use during the life cycle of buildings: A method. *Building and Environment* 32(4): 321–329.

Alcorn, A. (1998) *Embodied Energy Coefficients of Building Materials*. Centre for Building Performance Research, Victoria University of Wellington.

Alcorn, A. (2003) *Embodied Energy and CO₂ Coefficients for NZ Building Materials*. Centre for Building Performance Research, Victoria University of Wellington.

Alcorn, A. and Wood, P. (1998) *New Zealand Building Materials Embodied Energy Coefficients Database, Volume II – Coefficients*. Centre for Building Performance Research, Victoria University of Wellington.

Alcorn, J. A. and Haslam, P. J. (1996) The Embodied Energy of a Standard House – Then and Now. In: *Proceedings of the Embodied Energy: The Current State of Play Seminar* (G. Treloar, R. Fay and S. Tucker, eds), pp. 133–140. Deakin University, Australia.

Baines, J. T. and Peet, N. J. (1995) *1991 Input–Output Energy Analysis Coefficients*. Taylor Baines and Associates (for Centre for Building Performance Research).

Baird, G. and Chan, S. A. (1983) *Energy Cost of Houses and Light Construction Buildings* (Report No. 76), New Zealand Energy Research and Development Committee, University of Auckland.

Bonny, S. and Reynolds, M. (1988) *New Zealand Houses Today*. Weldon.

Breuer, D. (1988) *Energy and Comfort Performance Monitoring of Passive Solar Energy Efficient New Zealand Residences* (Report No. 172), New Zealand Energy Research and Development, University of Auckland.

Breuer, D. (1994) *Energy-Wise Design for the Sun: Residential Design Guidelines for New Zealand* Pacific Energy Design Limited (for Energy Efficiency and Conservation Authority, New Zealand). [1st edn 1985]

Buchanan, A. H. and Honey, B. G. (1994) Energy and carbon dioxide implications of building construction. *Energy and Buildings* 20: 205–217.

Building Industry Authority (1998) *The New Zealand Building Code Handbook and Approved Documents*. Scenario Communications (for Building Industry Authority, New Zealand).

Building Industry Authority (2000) *Approved Document for New Zealand Building Code Energy Efficiency Clause H*, 2nd edn. Scenario Communications (for Building Industry Authority, New Zealand).

Camilleri, M. J. (2000a) *Implications of Climate Change for the Construction Sector: Houses* (BRANZ Study Report No. 94). Building Research Association of New Zealand.

Camilleri, M. J. (2000b) *A Draft Climate Change Sustainability Index for Houses* (BRANZ Study Report No. 95). Building Research Association of New Zealand.

Centre for Housing Research (2004a) *Changes in the Structure of the New Zealand Housing Market: Executive Summary*. (Prepared by DTZ New Zealand.)

Centre for Housing Research (2004b) *Housing Costs and Affordability in New Zealand: Executive Summary*. (Prepared by DTZ New Zealand.)

Clark, S. J., Jones, M. and Page, I. C. (2005) *New Zealand 2005 House Condition Survey* (BRANZ Study Report No. 142). Building Research Association of New Zealand.

Culaba, A. B. and Purvis, M. R. I. (1999) A methodology for the life cycle and sustainability analysis of manufacturing processes. *Journal of Cleaner Production* 7: 435–445.

Fay, M. R. (1999) *Comparative Life Cycle Energy Studies of Typical Australian Suburban Dwellings.* PhD Thesis, Deakin University, Australia

Fay, R., Treloar, G. and Iyer-Raniga, U. (2000) Life-cycle analysis of buildings: a case study. *Building Research & Information* 28(1): 31–41

Geraghty, P. (1983) Environmental assessment and the application of expert systems: an overview. *Journal of Environmental Management* 39: 27–38

Honey, B. G. and Buchanan, A. H. (1992) *Environmental Impacts of the New Zealand Building Industry* (Research Report 92/2). Department of Civil Engineering, University of Canterbury, New Zealand.

Isaacs, N. P. and Trethowen, H. A. (1985) *A Survey of House Insulation* (Research Report R46). Building Research Association of New Zealand.

Isaacs, N. (2004) Supply Requires Demand – Where Does All of New Zealand's Energy Go? *Royal Society of New Zealand Conference*, Christchurch, 18 November 2004 (reprint *BRANZ Conference Paper No. 110, 2004*).

Jaques, R. (1996) Energy Efficiency Building Standards Project: Review of Embodied Energy. In: *Proceedings of Embodied Energy: The Current State of Play Seminar* (G. Treloar, R. Fay and S. Tucker, eds), pp. 7–14. Deakin University, Australia.

Johnstone, I. M. (2001) Energy use during the life cycle of single unit dwellings: Example. *Building and Environment* 32(4): 321–329.

Mercury Energy (2007) *Residential price plans.* Available at: http://www.mercury.co.nz/Residential/price_elec_vector.aspx#anytimestd [accessed 06 February 2007].

Ministry for the Environment (2000) *National Inventory Report: New Zealand Greenhouse Gas Inventory 1990–1998.* Ministry for the Environment.

Ministry for the Environment (2006) *New Zealand's Greenhouse Gas Inventory 1990–2004.* Ministry for the Environment.

Mithraratne, M. N. S. (2001) *Life Cycle Energy Requirements of Residential Buildings in New Zealand.* PhD thesis, The University of Auckland, New Zealand.

NZS4218:1996. *Energy Efficiency: Housing and Small Building Envelope.* Standards New Zealand.

Page, I. (1999) Comparing costs of timber with other structures. *Build* Nov–Dec 1999: 8–9.

Page, I. and Stoecklein, A. (1997) Development of Acceptable Solutions for the Revised Energy Clause of the New Zealand Building Code (Residential Buildings). *IPENZ Annual Conference*, Dunedin, New Zealand, February 1996 (reprint *BRANZ Conference Paper No. 32, 1997*).

Rawlinsons & Co. (1998) *Rawlinsons: New Zealand Construction Handbook.* Rawlinsons New Zealand Construction Handbook Ltd.

Shelton, R. (1999) Alternatives to timber framing: the structural issues. *Build* Nov–Dec 1999: 28–32.

Stoecklein, A., Pollard, A., Camilleri, M. et al. (2002) Findings from the Household Energy End-use Project (HEEP). *International Symposium on Highly Efficient Use of*

Energy and Reduction of its Environmental Impact, Osaka, January 2002 (reprint *BRANZ Conference Paper No. 102, 2002*)

Treloar, G. (1996) A Complete Model of Embodied Energy 'Pathways' for Residential Buildings. In: *Proceedings of the Embodied Energy: The Current State of Play Seminar* (G. Treloar, R. Fay and S. Tucker, eds), pp. 7–14. Deakin University, Australia.

Vale, B., Mithraratne, N. and Vale, R. (2000) Superinsulation for the Auckland Climate. In: *Architecture, City, Environment; Proceedings of PLEA 2000 Conference* (K. Steemers and S. Yannas, eds), pp. 150–154. James and James.

Wainwright, W. H. and Wood, A. A. B. (1981) *Practical Builder's Estimating*, 4th edn. Hutchinson & Co. Ltd.

Williams-Ellis, C., Eastcrich-Field, J. and Eastcrich-Field, E. (1947) *Building in Cob, Pisé and Stabilized Earth*, Revised edn. Country Life.

Wilson, D. (2006) *New Zealand Building Economist*. Plans & Specifications Ltd., 35(3).

Wooley, T., Kimmins, S., Harrison, P. and Harrison, R. (1997) *Green Building Handbook: A Guide to Building Products and Their Impact on the Environment*. E & F Spon.

Wright, J. and Baines, J. (1986) *Supply Curves of Conserved Energy: The Potential for Conservation in New Zealand's Houses*. Centre for Resource Management, University of Canterbury and Lincoln College, New Zealand.

7 Life cycle performance of an average New Zealand house

In order to test the model thus developed and to investigate the actual performance of average NZ houses, design and construction types and operating requirements representative of the current practices in New Zealand have to be identified, selected and simulated.

Average NZ house

New Zealand housing has undergone many changes over time. Although the typical suburban plot or section used to be around $1000\,m^2$, most city sections today are significantly smaller. During the late 1940s to 1950s housing was inexpensive and plain, and the main features included open-plan designs, simple shape with low-pitched roofs and outdoor decks through French doors. Housing constructed during the 1960s resulted in large subdivisions being filled with numerous houses that were very similar to each other, and social problems arose owing to the lack of communal facilities (Bonny and Reynolds 1988). The trend since the 1990s has been to build larger houses with three or more bedrooms.

There was a 37% increase in the total number of houses built in New Zealand between 1981 and 2001. In addition, the floor area of the average new house in New Zealand increased by 25% between 1970 and 2000, with about 55% of the current housing stock having being built since 1970 (CfHR[1], 2004a). The average New Zealand house with three bedrooms has a higher floor area in comparison with its British counterpart. The current trend in the New Zealand housing sector is, as already mentioned, to construct larger houses with three or more bedrooms. To meet the needs of today's complex society the function of the house is changing, leading to more energy intensive spaces catering for requirements of growing relationships, working from home, security, relaxation, entertainment, etc. Data on the current housing stock built since 1970 are given in Table 7.1.

Owing to the ethnic and cultural diversity of the present New Zealand population, especially in the Auckland region, extended family living has become more common among certain ethnic groups. This has not been catered for in the design of the

Table 7.1   Data on the current housing stock built since 1970

Year of construction	Per cent of total stock	Avg. floor area (m²)
1970s	18.7%	146
1980s	13.0%	149
1990s	13.2%	173
Since 2000	10.3%	194
Total	**55.2%**	–

average NZ house, which typically has comparatively small living rooms and other service areas apart from an inadequate number of bedrooms. However, the general trend in Auckland has been a reduction in the number of occupants per house. The affordability of housing has also decreased, particularly during the last 15 years, mainly due to significant appreciation of the value of land in certain parts of New Zealand (CfHR 2004b). The present New Zealand house could be broadly described as one that has extravagant glazing areas, that is large in scale and that is less elaborate in terms of detail. Terraces, decks and gardens play an important role.

New Zealand can be broadly divided into two climatic zones in terms of housing construction, i.e. a warm zone and a cool zone. For increased energy efficiency in the housing sector, the Energy Efficiency Building Code (NZS 4218:1996) requires higher insulation and performance standards in the cool zone, which roughly corresponds to the South Island. Due to this marked variation in the climate, the construction types used could be expected to be quite different. This investigation is limited to house types in the Auckland region in the warm zone where 34% of the New Zealand population live.

Common design used for the New Zealand house

The majority of New Zealand houses come under the category of separate houses with three bedrooms (see Chapter 1, Urban development and residential constructions). The Building Industry Advisory Council's (BIAC) standard NZ house as published by Baird and Chan (1983), is a three-bedroom detached house, which has been the basis model used in New Zealand energy studies by various researchers (Baird and Chan 1983; Wright and Baines 1986; Breuer 1988; Honey and Buchanan 1992; Alcorn and Haslam 1996; Jaques 1996). Following this tradition, the BIAC standard house, which is taken to be representative of the common NZ house, is used for this study rather than an ideal house. It was assumed to be located on a flat site. The floor plan of the BIAC house is shown in Figure 7.1.

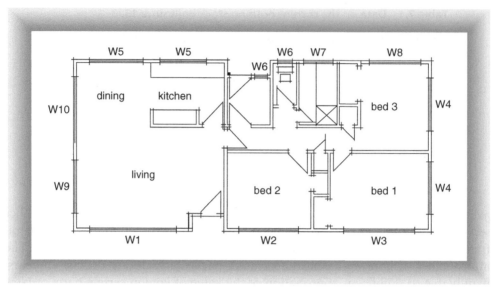

Fig. 7.1   Floor plan of BIAC standard house for New Zealand.
Based on: Baird, G. and Chan, S. A. (1983) *Energy Cost of Houses and Light Construction Buildings* (Report No. 76), New Zealand Energy Research and Development Committee, University of Auckland.

A brief description of the BIAC house is given in a list format, as follows.

- floor area 94 m² (14 m × 6.7 m);
- three bedrooms with open-plan living area, dining area and kitchen;
- separate bath/shower, WC, laundry;
- sloping ceiling with exposed rafters in the living and dining areas and flat ceiling to other areas;
- 12 lights and 16 power points.

However, as the average floor area of new houses in New Zealand has increased to 194 m² (see Table 7.1) the characteristics listed above would now represent the smaller and older houses of New Zealand. Wright and Baines (1986) used 119 m² as the floor area, being the average area of the NZ house at that time. Figure 7.2 shows the section and front elevation of the standard NZ BIAC house.

The published information for the BIAC house did not include a bill of quantities and, therefore, the calculations are based on the estimates of the major items/ material quantities and other information extracted from Baird's calculations. The following assumptions were used to facilitate the analysis.

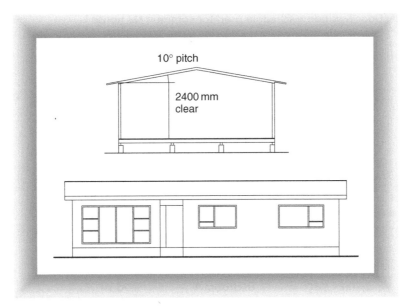

Fig. 7.2   Section and front elevation of BIAC standard house for New Zealand.
Based on: Baird, G. and Chan, S. A. (1983) *Energy Cost of Houses and Light Construction Buildings* (Report No. 76), New Zealand Energy Research and Development Committee, University of Auckland.

- The houses are located in Auckland.
- The Useful life of New Zealand houses is 100 years.
- No major refurbishment is carried out during the useful life other than the normal maintenance to maintain the houses at a habitable level.
- The embodied energy of New Zealand building materials and construction practices remains static over the useful-life time.

Foundations

Foundations for the light and superinsulated constructions were assumed as precast concrete piles of $200 \times 200 \times 600$ mm on $300 \times 300 \times 100$ mm concrete footings, to satisfy the requirements of NZS3604:1999.

A continuous reinforced strip footing of 300 mm (W) 200 mm (D) with 2nos. D12 and R6 @ 600 was used for the heavy construction type.

Doors and windows

No detailed study was done to determine the commonly used profiles for aluminium window sections and therefore the aluminium weight used in

Table 7.2   Door & window schedule used for common NZ house

Ref	Size (mm)	Location	Description	No.
D_1	2,055 × 860	Front	Standard fully glazed front door	01
D_2	2,055 × 890	Rear	Standard fully glazed door	01
D_3	1,980 × 810	internal door	paint quality MDF door	07
D_4	1,980 × 710	toilet	-do-	02
D_5	1,980 × 610	cupboard	-do-	02
W_1	3,400 × 2,050	Living	powder coated aluminium s/glazed	01
W_2	2,400 × 1,300	Bedroom 2	-do-	01
W_3	2,700 × 1,300	Bedroom 1	-do-	01
W_4	2,000 × 700	Bedrooms 1&3	-do-	02
W_5	2,100 × 850	Dining/kitchen	-do-	02
W_6	600 × 850	Toilet/laundry	-do-	02
W_7	1,100 × 850	Bath	-do-	01
W_8	2,000 × 1,300	Bedroom 3	-do-	01
W_9	2,000 × 1,500	Living	-do-	01
W_{10}	2,500 × 1,500	Dining	-do-	01

Baird's study was used here. The schedule of doors and windows used for the study (Table 7.2) was derived from the drawings published. Single glazing is commonly used in NZ houses and, therefore, to satisfy the requirements of the energy efficiency code (NZS4218:1996), the area of openings on any external wall of the BIAC house was limited to 30% of the external wall area.

Electrical services

According to Fay (1999), electrical services represent only 1% of the total embodied energy of houses irrespective of the house type. The electrical service requirements of the typical NZ house were assumed as follows based on Fay (1999).

Table 7.3 Material composition for common electrical fittings

Item	Material	Mass
Switchboard	Steel	4 kg
	Copper	1 kg
	Plastic	2 kg
Power cable	Plastic	0.135 kg/m
	Copper	0.065 kg/m
Lighting cable	Plastic	0.10 kg/m
	Copper	0.03 kg/m
Television cable	Plastic	0.05 kg/m
	Copper	0.03 kg/m
Power outlets	Plastic	0.1 kg
	Steel	0.1 kg
Light switch	Plastic	0.04 kg
	Copper	0.01 kg
Ceiling rose	Plastic	0.03 kg
	Copper	0.04 kg
Light fitting	Plastic	0.2 kg
	Steel	0.2 kg
	Glass	–
	Copper	0.1 kg

Source: Fay, M. R. (1999) *Comparative Life Cycle Energy Studies of Typical Australian Suburban Dwellings.* PhD Thesis, Deakin University, Australia.

Electrical cables	Mains connection	20 m power cable
	Power outlet	10 m power cable
	Light point	15 m lighting cable
Fittings	Material compositions for common electrical fittings were assumed as per Table 7.3.	

Plumbing and sanitary fittings

The data found in Table 7.4 were used for plumbing services, taken to be representative of current construction practices in New Zealand based on previous studies by Fay (1999) and Baird and Chan (1983).

Table 7.4   Schedule of plumbing, drainage and sanitary fittings

Item	Material	Mass	Comments
125 mm PVC spouting (i.e. guttering)	Plastic	0.65 kg/m	estimated
65 mm PVC downpipes	Plastic	0.65 kg/m	estimated
25 mm copper hot and cold water pipes	Copper	0.829 kg/m	Fay
100 mm PVC waste & storm water pipes	Plastic	1.0181 kg/m	calculated
Bath 1,675 mm pressed steel white enamel	Steel	55 kg	Fay
Shower 925 × 925 × 100 mm	Stainless steel	20 kg	assumed
Vanity basin 580 × 415 mm	Vitreous china	12 kg	Fay
Sink units 1,830 mm	Stainless steel	10 kg	assumed
WC pan	Vitreous china	17 kg	Fay
Double flap seat, cistern and accessories	Plastic	10 kg	Fay
Mains pressure 135 litres hot water cylinder	Steel	14 kg	Baird
	Copper	16 kg	
Taps and valves 20 mm bibcock 2 nos. each for kitchen, laundry, shower, bath and vanity	Steel	1.5 kg/set	assumed
20 mm hose taps	Brass	1.5 kg	assumed
25 mm stop valve	Brass	1.5 kg	assumed
Toilet paper holder	Steel	0.1 kg	assumed
Towel rail	Steel	0.2 kg	assumed

Domestic appliances and furniture

The domestic appliances and furniture used in the typical NZ house are as shown in Tables 6.5 and 6.6 respectively.

Operating energy requirements

Space heating energy

The space heating energy requirement for this common NZ house was calculated using the ALF 3.0 (annual loss factor) thermal simulation software developed by BRANZ. In order to accurately represent the current usage of houses, the following two heating regimes were modelled.

1. All day heating (7.00 to 23.00 hrs) – to represent families with smaller children and those who work from home; and
2. Intermittent heating (7.00 to 9.00 hrs in the morning and 17.00 to 23.00 hrs in the evening) – to represent families with both adults working away from home.

It was assumed an internal temperature of 18°C would be maintained in all houses. Annual heating energy requirements for the six generic construction types (light construction, heavy construction, superinsulated construction, timber-framed concrete floor construction, mud brick wall construction and interlocking solid timber construction) calculated according to the above requirements are as shown in Table 7.5.

Electrical lighting

Electrical lighting usages in the average NZ house, based on the figures identified in Table 6.8 and the floor plan and the specifications published by the Baird and Chan (1983: p. 28) study, are as shown in Table 7.6.

Based on the above usage, the total electrical energy requirement for 12 lights is 863 kWh per annum. Average electricity usage for lighting according to the HEEP study (Isaacs 2004) is 1185 kWh per annum. The level of 863 kWh per annum for the smaller NZ house considered here would therefore be representative of the situation.

Table 7.5   Space heating energy requirements for a typical NZ house in Auckland

Construction type	Heating schedule	Energy requirement (kWh per annum)
Light construction	All day	2,123
	Intermittent	1,768
Heavy construction	All day	2,019
	Intermittent	1,784
Superinsulated construction	All day	1,115
	Intermittent	591
Light construction with concrete floor	All day	2,005
	Intermittent	1,770
Mud brick wall with concrete floor construction	All day	2,436
	Intermittent	2,176
Interlocking solid timber construction	All day	2,479
	Intermittent	2,110

Table 7.6   Electrical energy requirements for lighting of the average NZ house

Location	Wattage	Hours/day	Useful life (@ 1000 h)	kWh per annum
Living/dining	3×60	4.0	8 months	262.80
Kitchen	1×100	6.0	5.5 months	219.00
Bedroom 1	1×100	0.5	5 years 6 months	18.25
Bedrooms 2 & 3	2×75	0.5	5 years 6 months	27.38
Hallway	1×100	2.5	1 year 2 months	91.25
Bathroom/toilet	2×75	0.75	3 years 8 months	41.06
Laundry	1×75	0.75	3 years 8 months	20.53
External access	1×100	5.0	6 months	182.50

Appliances

The total electrical energy requirement for operation of domestic appliances is calculated based on the data shown in Table 6.8. The total electricity consumption for domestic appliances thus calculated amounts to 2627 kWh per annum.

Water heating and cooking

The total electricity usages for water heating and cooking are based on the HEEP results and are 4000 kWh per annum and 474 kWh per annum respectively.

This model was used for life cycle analysis of typical NZ houses based on life cycle energy, cost and environmental impact, and the initial investigations were used to identify:

- the most effective construction type; and
- the effect of using higher levels of insulation in the typical NZ house.

The results of these analyses are discussed next.

Life cycle energy analysis of an NZ house – model results

The direct and indirect energy required to build a standard NZ house were estimated using the life cycle analysis model developed as discussed in the previous chapter. Initial embodied energy requirements for the average NZ house based on the most common construction types and the superinsulated construction are summarised in Table 7.7.

Table 7.7 Initial embodied energy intensities for BIAC standard house

Building element	Embodied energy intensity (MJ/m²)					
	Light construction		Heavy construction		Superinsulated construction	
Foundation	30	2%	80	3%	30	2%
Floor	220	12%	740	28%	370	16%
Walls	450	25%	820	31%	580	25%
Roof	400	22%	340	13%	560	24%
Joinery	230	13%	230	9%	300	13%
Electrical work	100	5%	100	4%	100	4%
Plumbing	170	9%	170	6%	170	7%
Finishes	210	12%	160	6%	200	9%
Total	**1,810**	**100%**	**2,640**	**100%**	**2,310**	**100%**

* Due to rounding off, components may not add up to the total figure

For all three construction types (light, heavy and superinsulated), floor, walls and roof collectively represent the bulk (59%, 72% and 65%) of the initial embodied energy of the New Zealand house. From the above it could be concluded that preliminary energy calculations for just the main building elements, i.e. floor, walls and roof, could aid in the selection of design and construction types appropriate for any situation.

The floor construction commonly used in New Zealand is aluminium foil insulation draped over the timber floor framing. Theoretically, this construction would provide an R value of $1.4 \, \text{m}^2\text{C}^0/\text{W}$, although practically this may not be achieved as found by previous research (discussed in Chapter 6). The floor construction used for the superinsulated construction of 200 mm of glass fibre insulation with a 3 mm plywood layer fixed to the underside of the floor joists would provide an R value of $4.4 \, \text{m}^2\text{C}^0/\text{W}$. However, according to the above life cycle embodied energy comparison, the superinsulated construction has a 68% higher initial embodied energy for floor construction compared with that for the light construction. Similarly, wall and roof constructions used for superinsulated construction are 29% and 40% higher in initial embodied energy compared to that of the light construction, while double glazing is 30% higher in initial embodied energy compared with single glazing.

Life cycle embodied energy for light, heavy and superinsulated construction types calculated based on the best estimate for replacement cycles (as per

Table 6.4) was 4274, 4109 and 4857 MJ/m², respectively. Heavy construction showed about a 4% reduction in life cycle embodied energy compared with that of light construction, mainly due to a reduced maintenance requirement over the useful life. A comparison of life cycle embodied energy for light, heavy and superinsulated construction types and their composition is given in Figure 7.3.

These estimates are approximate and do not include energy for transport of materials to the site, on-site requirements or material wastage. These values are, however, not as critical as those for the energy embodied in the manufacture of building materials. Transport energy and the on-site requirements have been estimated by Baird and Chan (1983, p. 8) to be 6% of the total gross energy requirement. Once transport and site energy components are included, life cycle embodied energy of light, heavy and insulated constructions increase to 4531, 4356 and 5148 MJ/m², respectively.

Life cycle energy requirements of the common construction types were then compared to evaluate the most effective construction type in life cycle energy terms. Only the space heating component of operating energy is dependent on design and construction, while the balance depends on user behaviour, and therefore this comparison only considered embodied and space heating energy requirements. Figure 7.4 shows a comparison of embodied and space heating energy (with all-day heating) for common construction types used in New Zealand. Although this does not include embodied energy of furniture and appliances or operating energy other than space heating energy, 6% of the construction embodied energy has been added to account for transport and site energy components. Comparison of embodied and space heating energy for common construction types used in New Zealand with intermittent heating is as shown in Figure 7.5.

The space heating energy requirement of the superinsulated construction with both heating schedules is markedly low compared with that of the common construction, being 48% and 67% lower with all-day and intermittent heating schedules, respectively. However, with the heavy construction type, space heating energy requirement is 5% lower than for the common construction with an all-day heating schedule while it is 1% higher with the intermittent heating schedule. Owing to the need for mass heat-up, the heavy construction type uses marginally larger amounts of energy with an intermittent heating schedule.

In terms of life cycle energy however, both heavy and superinsulated construction types perform better than the light construction type irrespective of the space heating schedule used. Life cycle energy demands of heavy and superinsulated

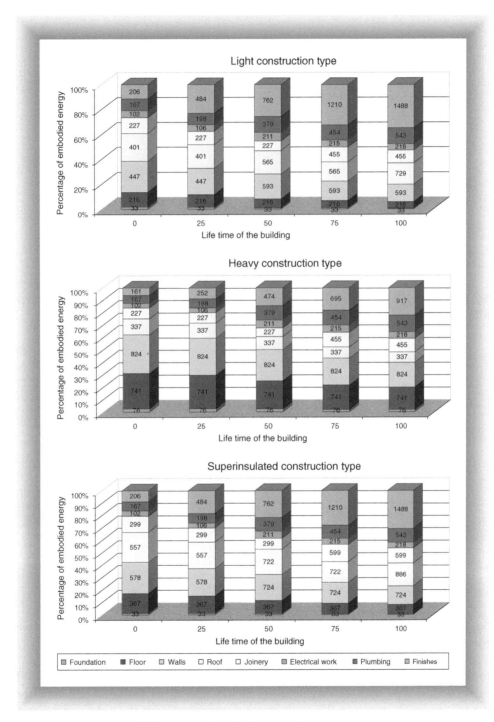

Fig. 7.3 Comparison of life cycle embodied energy for common construction types.

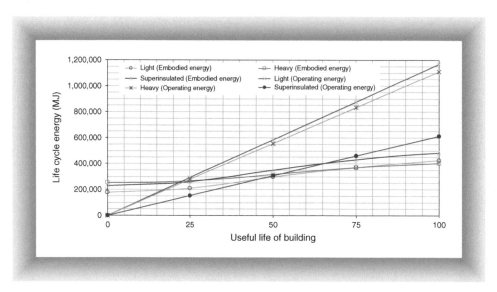

Fig. 7.4 Life cycle operating and embodied energy use for common construction types with all-day heating (appliances and furniture excluded).

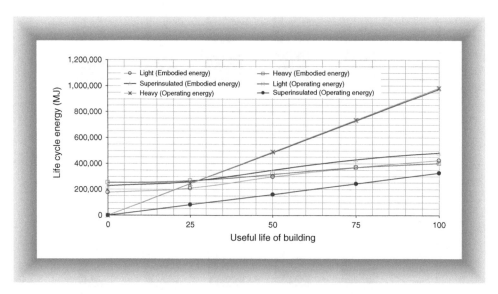

Fig. 7.5 Life cycle operating and embodied energy use for common construction types with intermittent heating (appliances and furniture excluded).

constructions are respectively 5% and 31% less than that of the common light construction type with all-day heating. Heavy and superinsulated constructions are 1% and 67% lower in life cycle energy compared with the light construction with intermittent heating. Although the space heating energy requirement of the heavy construction type is marginally higher than that of the light construction type with the intermittent heating schedule, due to low maintenance requirements of the heavy construction type life cycle energy of heavy construction is lower than that of light construction. As this indicates that the relative performances of the construction types are not significantly affected by the space heating schedule used, the rest of the analysis is based on all-day heating for 18°C. (The impacts of different heating schedules are analysed later in this section.)

According to Baines and Peet (1995), cited by Alcorn (1996), in New Zealand generation of 1 MJ of electricity requires 1.53 MJ of primary energy. Therefore, operating energy requirements were converted to primary energy using this relationship and with the assumption that electricity is the main source of space heating energy. According to the HEEP study (Isaacs et al. 2006), average space heating energy use in an NZ house is 34% of the total operating energy. However, one could argue that the omission of operating energy other than space heating ignores the potential of efficiency improvement measures such as fuel switching and additional insulation, to reduce the energy use for water heating, which amounts to 29% of the total according to the same study. Table 7.8 gives data on life cycle energy requirements of common construction types used in New Zealand with all-day heating and Figure 7.6 is a graphical representation thereof.

Figure 7.7 shows the total life cycle energy usage including that for domestic appliances and furniture. This includes life cycle embodied energy and operating energy for the building and the appliances and furniture. For the light construction type, 21% of the operating energy is for space heating while a further 40% and 26% relates to water heating and operation of appliances respectively. The balance of 13% corresponds to lighting and cooking requirements. While 18% of the total life cycle operating energy is used for space heating with the heavy construction type, a further 41% and 27% are used for water heating and for operation of appliances respectively. For the superinsulated construction type, space heating energy is only 7% of the total operating energy, water heating and operation of appliances 47% and 31% respectively. However, only the space heating component of operating energy is dependent on design and construction, while the balance depends on user behaviour.

Although the embodied energy of heavy construction is higher (46% higher than that of light construction) at the initial stage, it is about 4% lower at the end of the useful life. This could be expected due to the lower maintenance requirements of the heavy construction. Embodied energy of superinsulated construction remains higher than that of light construction throughout the lifetime, at 26% at the initial stage and 11% at the end of the useful lifetime. However, embodied energy is only 27% of the total life cycle energy for both light and heavy

Table 7.8 Life cycle energy requirements for common construction types (excluding furniture and appliances)

Category	Life cycle energy intensity (MJ/m^2)				
	Year 0	Year 25	Year 50	Year 75	Year 100
Light construction type					
Construction embodied energy	1,800	2,112	2,986	3,740	4,274
Transport & site energy (6% of construction energy)	108	127	179	224	257
Total embodied energy	1,908	2,239	3,165	3,964	4,531
Space heating (2,123 kWh per annum)	0	3,110	6,220	9,330	12,440
Total	1,908	5,349	9,385	13,294	16.971
Heavy construction type					
Construction embodied energy	2,635	2,760	3,268	3,795	4,109
Transport & site energy (6% of construction energy)	158	166	196	228	247
Total embodied energy	2,793	2,925	3,464	4,023	4,356
Space heating (2,019 kWh per annum)	0	2,958	5,915	8,873	11,830
Total	2,793	5,883	9,379	12,896	16,186
Superinsulated construction type					
Construction embodied energy	2,310	2,623	3,496	4,322	4,857
Transport & site energy (6% of construction energy)	139	157	210	259	291
Total embodied energy	2,448	2,780	3,706	4,581	5,148
Space heating (1,115 kWh per annum)	0	1,039	2,077	3,116	4,154
Total	2,448	4,413	6,973	9,481	11,681

construction types while it is 44% of the total life cycle energy for the super-insulated construction type, at the end of the useful life. This is comparable to 24% reported by Jaques for an average NZ house in his research (1996). On the other hand, as indicated by Table 7.1, the floor area of the average NZ house is rapidly increasing, leading to a higher embodied energy component than that of the house used for this study. (The impact of house size on life cycle energy is

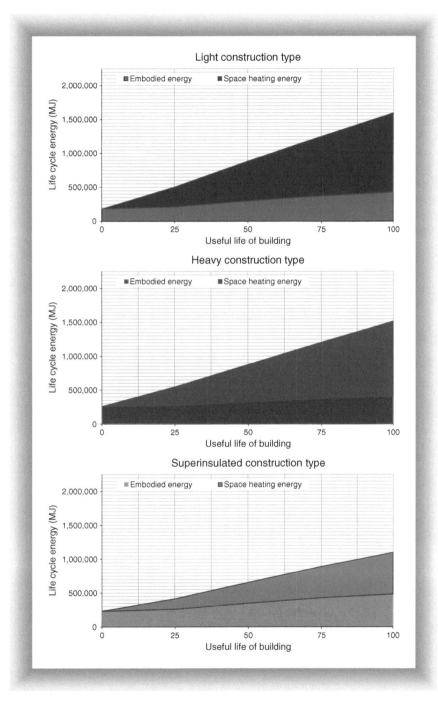

Fig. 7.6 Comparison of life cycle energy for common construction types with all-day heating (furniture and appliances excluded).

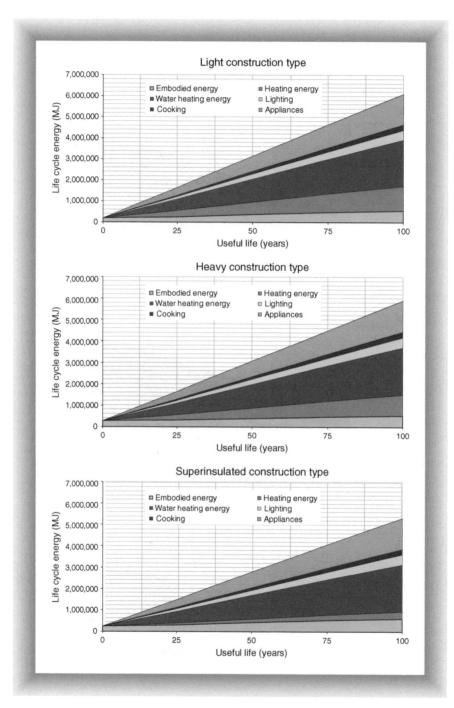

Fig. 7.7   Comparison of life cycle energy for common construction types (furniture and appliances included).

analysed later in this section.) However, operating energy is the crucial component of life cycle energy. The life cycle operating energy requirement of heavy and superinsulated constructions was respectively 3% and 15% less than that of the light construction type at the end of the useful life, when appliances and furniture are included in the analysis.

Since operating energy is the dominant component in life cycle energy, heavy construction with higher thermal mass and, therefore, potentially less operating energy might be expected to have lower life cycle energy. However, the reduction in life cycle energy was not significant, with heavy construction being 1% higher at 25 years and only 2%, 3% and 3% lower in comparison with the light construction type at years 50, 75 and 100, respectively. However, it should be noted that the BIAC house has not been designed to optimise the use of the mass or to combine it with increased insulation.

In contrast, the superinsulated house with its relatively low space heating energy reduced the life cycle energy by 13% over a 100 year lifetime. At the 25th year, life cycle energy of superinsulated construction was 10% lower than that of the common light construction type while it was 12% lower at the 50th and 75th years. Table 7.9 gives a comparison of life cycle energy for all six generic construction types identified by the initial study discussed earlier (see Chapter 6, Establishing the knowledge base of the model).

The comparison given in Table 7.9 does not include operating energy, with the exception of space heating energy. This component would be the same for all the construction types considered and therefore could be omitted from the analysis. Figure 7.7 shows the impact of this component on the life cycle energy for the three basic construction types considered. However, the comparison made shows that, for a standard non-optimised house, superinsulated construction had the lowest life cycle energy with the fourth highest initial embodied energy. In contrast, solid timber construction, with the second-lowest initial embodied energy, had a very high life cycle energy, 45% higher than that of a superinsulated construction. The timber construction with concrete floor, with a marginally lower initial embodied energy compared with superinsulated construction, is second-best in terms of life cycle performance. A comparison of life cycle energy for the six main construction types is shown in Figure 7.8.

Although the BIAC standard house has a total floor area of $94\,m^2$, over the years the floor area of new houses constructed in NZ has increased, as shown in Table 7.1. These larger new houses now represent 55% of the existing housing stock. In order to investigate the influence of increased floor area on the life cycle energy, the BIAC house was then enlarged to $146\,m^2$ and $194\,m^2$ (to represent

Table 7.9   Comparison of life cycle energy for different construction types

Category	Life cycle energy intensity (MJ/m^2)				
	Year 0	Year 25	Year 50	Year 75	Year 100
Light construction type					
Total embodied energy	1,908	2,239	3,165	3,964	4,531
Space heating (2,123 kWh per annum)	0	3,110	6,220	9,330	12,440
Total	1,908	5,349	9,385	13,294	16,971
Heavy construction type					
Total embodied energy	2,793	2,925	3,464	4,023	4,356
Space heating (2,019 kWh per annum)	0	2,958	5,915	8,873	11,830
Total	2,793	5,883	9,379	12,896	16,186
Superinsulated construction type					
Total embodied energy	2,448	2,780	3,706	4,581	5,148
Space heating (1,115 kWh per annum)	0	1,633	3,267	4,900	6,533
Total	2,448	4,413	6,973	9,481	11,681
Light construction with concrete floor type					
Total embodied energy	2,441	2,590	3,469	4,044	4,566
Space heating (2,005 kWh per annum)	0	2,578	5,156	7,734	10,313
Total	2,441	5,168	8,625	11,778	14,879
Earth-brick construction type					
Total embodied energy	2,652	2,784	3,323	3,882	4,215
Space heating (2,436 kWh per annum)	0	3,132	6,265	9,397	12,529
Total	2,652	5,917	9,587	13,279	16,744
Solid timber construction type					
Total embodied energy	1,936	2,097	3,344	3,826	4,256
Space heating (2,479 kWh per annum)	0	3,188	6,375	9,563	12,750
Total	1,936	5,285	9,719	13,389	17,006

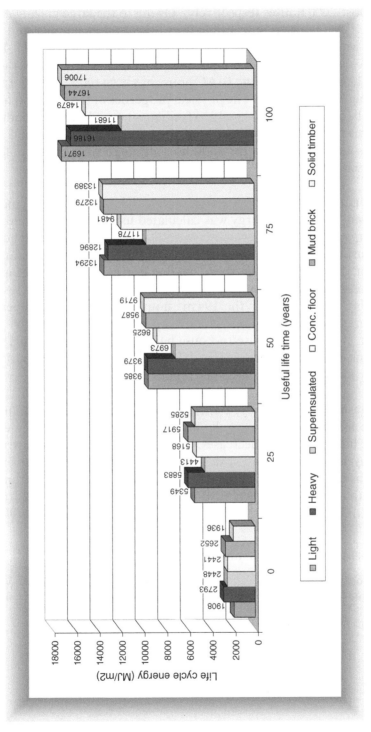

Fig. 7.8 Comparison of life cycle energy for different construction types.

1970s and post-2000 houses respectively). Owing to the large glazing areas used, the generic BIAC house is susceptible to overheating in summer and high heat loss during winter. Hence, in keeping with the requirements of NZS 4218:1996, the window areas of the larger houses have been limited to 30% of the external wall area. The aspect ratio of the BIAC house has been retained for larger houses for a meaningful comparison. The physical properties of the common house were assumed to be as shown in Table 7.10 to facilitate the analysis of embodied energy of the houses. It was also assumed that the nature of electrical and plumbing installations would be similar for all houses although wiring and roof guttering were increased accordingly.

Two constructions, common light construction and superinsulated construction, were analysed. Embodied energy was calculated using the model, and the space heating energy demands for 146 and 194 m^2 light and superinsulated houses were calculated using ALF, based on the originally used requirements and all-day heating. For the 146 m^2 house, space heating energy use with light and super-insulated construction types was 3016 and 1577 kWh per annum, respectively. Space heating energy use for the 194 m^2 house with light and superinsulated construction types was 3862 and 2046 kWh per annum, respectively. However, this exercise, if done on a house-to-house basis as a comparison on a per metre squared basis, could disguise the impact of the bigger house. Comparison of life cycle energy for all six houses is as per Figure 7.9.

With superinsulation, the life cycle energy of the 94 m^2 house is 28% higher than that for the light construction at the initial stage. However, by the end of the lifetime, the life cycle energy decreases, compared with the light construction, by 31%. Life cycle energy increases by 37% and 39% for light and superinsulated constructions respectively when the floor area is increased by 55% from 94 m^2 to 146 m^2. On the other hand, when the floor area is increased by 106% (from 94 m^2 to 194 m^2), the life cycle energy of light construction increases by 79% but the

Table 7.10   Physical properties of the average NZ House

	Floor area (m^2)	Overall dimensions (m)	External wall area (m^2)	Internal wall area (m^2)	Window area (m^2)
BIAC house	94	6.7 × 14	93	211	30
1970s house	146	8.6 × 17	115	264	35
2000 house	194	9.7 × 20	133	304	40

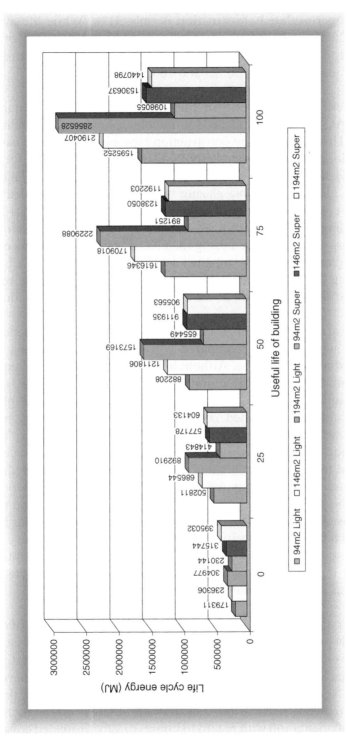

Fig. 7.9 Comparison of life cycle energy for houses of different sizes (furniture and appliances excluded).

increase for superinsulated construction is only 31%. While life cycle energy increases by 30% at the initial stage when a light construction of $194\,m^2$ is superinsulated, it is reduced by 32%, 42%, 47% and 50% compared with the light construction at years 25, 50, 75 and 100, respectively.

In order to investigate the influence of different climatic conditions on the life cycle energy, the BIAC house was modelled for the colder climate in Christchurch, in the South Island. However, the light construction type used in this analysis of houses in Auckland does not meet the code (NZS 4218:1996) requirement for the colder climate zone. Hence, a further construction type with additional ceiling insulation (120 mm fibre glass) and double glazing was also added to the analysis. Three constructions, common light construction, double glazed light construction with additional ceiling insulation and superinsulated construction, were analysed. Embodied energy was calculated using the model. The space heating energies for light, double glazed light and superinsulated houses located in Christchurch, calculated using ALF, were 6068, 4465 and 2981 kWh per annum, respectively. Figure 7.10 shows a comparison of life cycle energy for the two locations.

While in the warmer climate of Auckland, superinsulated construction reduced the life cycle energy by 31% compared with light construction, in the colder climate this reduction was 44%. Although the life cycle energy increased due to the higher space heating energy requirement in the colder climate, the pattern remains the same. In terms of initial embodied energy, double glazed light construction and superinsulated construction are 7% and 28% respectively higher than the corresponding value for the light construction. The use in Auckland of the double glazed light construction with additional ceiling insulation specified for the colder climatic zone of New Zealand, can reduce the life cycle energy use by 18% compared with the light construction type currently being used. This suggests that for a standard house of non-optimal design (the BIAC house used for this analysis was not designed as a passive solar low-energy house), in which form the majority of the new constructions are found, the use of light construction with increased ceiling insulation and double glazing, as specified for colder regions throughout New Zealand, could demonstrate significant savings in terms of life cycle energy.

A study (Mithraratne and Vale 2004), which considered the impact of orientation on the life cycle performance of common houses in New Zealand, found that it is beneficial to avoid south-east through east to north-east orientations for the main living areas, and to use north through to west orientations whenever possible everywhere in New Zealand. However, orientation did not affect the performance of superinsulated constructions in Auckland and, therefore, it is unnecessary to take account of orientation with regard to houses of superinsulated construction in

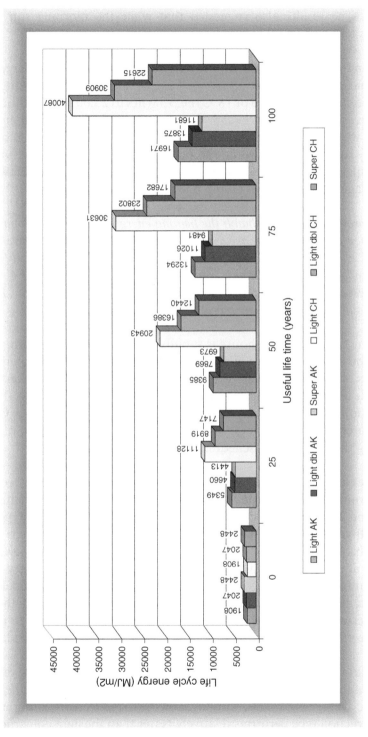

Fig. 7.10 Comparison of life cycle energy for different climates (furniture and appliances excluded).

that region. The results also suggested that while the morning sun is not of assistance in terms of useful solar energy, the afternoon sun is.

Space heating energy requirements for the average house considered here of various forms of construction, are based on heating of house throughout the day to maintain a temperature of 18°C (i.e. 7.00 hrs to 23.00 hrs). Although this is representative of families with children and those working from home, this assumption would seem somewhat unrealistic in the present context, most houses being empty during the day while the occupants work elsewhere. Further, the internal temperatures of New Zealand houses remained unchanged between 1971 and 1997. Page and Stoecklein (1997: p. 8) argue that:

> *As an approximate rule of thumb the heating energy use is proportional to the average temperature difference between internal and external air. In the moderate climate in most New Zealand population centres the difference between indoor and mean winter outdoor temperature is about 10°C or less ... an increase of mean internal temperatures by 2°C increases the heating energy requirement by 20% or more.*

However, the function of the house is rapidly changing and houses are modelled over a period of 100 years, while research suggests that the thermal comfort expectations of New Zealanders will improve over the years. To investigate the impact of the use of varying heating schedules on the life cycle energy, three types of construction with both all-day and intermittent heating to 18°C (see Operating energy requirements above for details) were compared. Comparison of life cycle energy using the two heating schedules is as shown in Figure 7.11.

The life cycle energy decreased owing to the lower space heating energy requirement with an intermittent heating schedule, for all construction types. The reduction was 12%, 9% and 26% for light, heavy and superinsulated construction types respectively at 100 years. With all-day heating, the heavy construction type performs marginally better (5% less) than the light construction type in life cycle energy terms. In contrast, with intermittent heating, the performance of heavy construction is similar to that of light construction in life cycle energy terms. Superinsulated construction, however, reduced life cycle energy use irrespective of the heating schedule selected. Therefore, the selection of heavy construction has to match the intended use of the house, which is likely to change over the life of the house.

Although life cycle energy varies with the heating schedule, the ranking remains the same. Therefore, operating energy is the dominant component of the life cycle energy of the average NZ house. Although, owing to reduced operating energy, the heavy construction type performs relatively better than the light construction type

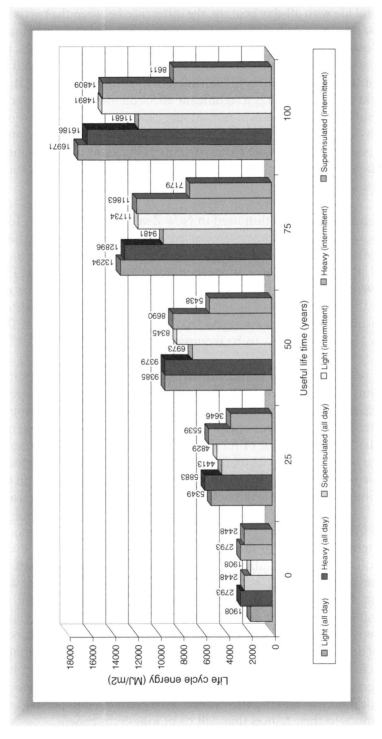

Fig. 7.11 Comparison of life cycle energy with different heating schedules (furniture and appliances excluded).

over the useful life with all-day heating, the difference between the two is not great. Therefore, at the lower end of the housing market the common light construction, which has lower initial embodied energy and lower initial cost, does provide a reasonably 'good deal' in life cycle terms. However, since superinsulated construction reduced the life cycle energy, the simple act of increasing insulation in the standard light construction has a significant effect on improving the performance over the useful life. Reduction in life cycle energy is not reliant on the use of thermal mass, which is a less common construction type in New Zealand due to the requirements of the Earthquake Code. Although bigger houses with higher floor areas and those located in colder climates use more life cycle energy, superinsulation could be used here as well to reduce the overall impact.

The common floor construction used in New Zealand of aluminium foil draped over the floor framing has a lower embodied energy compared with that of glass fibre under-floor insulation, although draped foil has been proven to be associated with practical problems of proper installation (Isaacs and Trethowen 1985). However, the glass fibre under-floor insulation reduced the operating energy and therefore operating costs and could be considered a better option for floor insulation. This may suggest that the standard practice of floor construction in NZ houses needs to be reconsidered.

Once space heating energy is reduced, alternative measures to reduce other operating energy usages such as for water heating, appliances and lighting, are essential. The next step in performance improvements, therefore, would be the introduction of solar water heating, efficient appliances and compact fluorescent lighting. The high embodied energy content of furniture which needs regular replacement over the life of the house owing to its relatively shorter useful life can be reduced by the use of recycled (i.e. antique or second hand) furniture. The construction of smaller houses is another alternative approach to reducing embodied energy use.

A reduction in life cycle energy use of houses would (presumably) be good for New Zealand in terms of its Kyoto Protocol commitments. Although life cycle energy could be reduced by using additional insulation, the decision to invest or not to invest in insulation would depend on the cost. As discussed in Chapter 4, research suggests that New Zealand designers perceive the client requirement to be for minimum initial cost and not minimum life cycle cost. This is further aggravated by the current practice of changing ownership of NZ houses, which occurs at a frequency of approximately every 7 years. However, the initial cost is about 50% of total life cycle cost for most building types and life cycle costs could be used to inform designers and clients of the total cost implications of their decisions. On the national scale it may be that incentives need to be considered to persuade people to make alterations to their houses that will be of long-term benefit.

Life cycle cost analysis of NZ house – model results

As discussed in Chapter 4, the results of life cycle cost analyses depend on the discount rate selected, how inflation is taken into account, the period of analysis used, base date or date of commencement of the analysis, and types of costs included and/or ignored together with the method of analysis used.

Since the useful life of NZ houses is assumed to be 100 years, the method of analysis employed should be a discounting method, and the period of analysis 100 years. The beginning of occupancy is used as the base date or the beginning of the analysis, based on the common practice used for building life cycle costing (LCC) analysis. Although 3–4% is used as the discount rate for international studies, a 5% discount rate was selected based on the recommendations of the Australia/New Zealand Standard. Discounted real costs were used for the analysis and therefore do not include inflation. However, use of real costs provides an accurate comparison due to use of current values, and the need to predict future inflation is eliminated. The net present value (NPV) method was selected for this analysis as NPV would represent the present value of the total investment required over the useful life to maintain different generic constructions used in NZ houses. This is the amount that has to be set aside today to cover all the costs incurred throughout the useful life. Future costs incurred over the useful life are discounted from the date on which they occur back to the beginning of occupation and then added to produce the NPV of the life cycle cost.

Life cycle cost was calculated for the three constructions on the same basis as for life cycle energy, i.e. without taking appliances and furniture into account. The results are as shown in Table 7.11.

While the constructions considered above are in the low-cost range in the market, the costs involved relate to building works only and activities such as preliminaries and site works prior to construction etc. are not included. The latter are considered to be the same for all the constructions considered and therefore are omitted from the analysis. While no GST is added to the initial construction cost, 12.5% has been added to replacement work as GST. According to New Zealand Building Economist (Wilson 2006), the average cost of construction in Auckland is about 1089NZ$/m^2 for a standard house, 1345NZ$/m^2 for an executive house and 1500NZ$/m^2 for an individual architect designed house. Current residential electricity price plans consist of two components: a line charge and a unit charge. As only space heating energy use is considered in the above analysis, the line charge (which represents the cost of being connected and is common to all electricity uses in the house) has not been included.

Table 7.11 Comparison of life cycle cost for different construction types

Category	Life cycle cost (NZ$/m^2)				
	Year 0	Year 25	Year 50	Year 75	Year 100
Light construction type					
Cost of construction	973	1,100	1,207	1,224	1,228
Cost of space heating energy	0	57	74	79	80
Total	973	1,157	1,281	1,303	1,308
Heavy construction type					
Cost of construction	1,177	1,292	1,374	1,395	1,397
Cost of Space heating energy	0	54	70	75	76
Total	1,177	1,346	1,444	1,470	1,473
Superinsulated construction type					
Cost of construction	1,148	1,301	1,381	1,407	1,410
Cost of space heating energy	0	30	39	41	42
Total	1,148	1,331	1.420	1,448	1,452

Both superinsulated and heavy constructions perform similarly in life cycle cost terms according to this analysis. While the superinsulated construction is approximately 11% greater in terms of life cycle cost than the light construction, with an approximate 18% increase in initial cost, the heavy construction type is 13% more expensive in life cycle terms and 21% more expensive at the initial stage. Figure 7.12 shows a comparison of life cycle cost and life cycle energy for various construction types without the inclusion of appliances and furniture.

While the life cycle costs increase until about year 60 and then level off, the energy savings continue over the entire life of the building. Although both heavy and superinsulated constructions perform similarly in terms of life cycle cost, superinsulated construction provides an approximate 31% reduction in life cycle energy in contrast to the 5% reduction that can be achieved by using heavy construction. Any additional cost might easily vanish if the fuel prices rise erratically over the next 100 years.

Although the operating energy uses for water heating, appliances and lighting are not dependent on construction type, these were then included in the analysis in order to obtain a complete picture of performance. Maintenance and replacement schedules for all furniture and appliances were also included. Since all

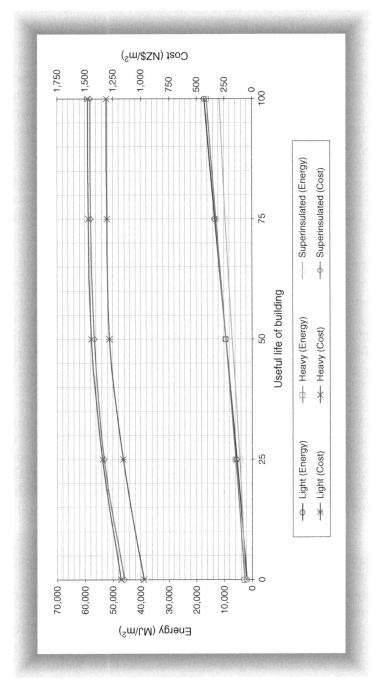

Fig. 7.12 Comparison of life cycle cost and life cycle energy for various constructions (furniture and appliances excluded).

operating energy uses are included in this analysis, electricity line charge was also taken into account. The costs of all these items and the replacement costs were also modelled. Figure 7.13 and Table 7.12 show the results when these requirements are incorporated.

When appliances are included in the analysis, the same rankings in terms of life cycle energy use still apply to the constructions. The superinsulated construction is now 13% lower in terms of life cycle energy use than the lightweight construction, with about 7% higher life cycle cost and 16% higher initial cost. Heavy construction, which is similar to superinsulated construction in terms of life cycle cost, is 18% more expensive than the light construction at the initial stage, and shows only a 3% reduction in life cycle energy over the light construction. This suggests that if mass is used in NZ houses it needs always to be used in conjunction with passive solar design, properly insulated and placed optimally in the building.

The marginal increase in the initial cost associated with higher insulation does not seem to provide benefit to the individual houseowner whether they change house in the short term or not. However, this additional insulation could buffer the owner against any sudden increase in energy prices, while maintaining the house at a higher internal temperature, which will offer improved comfort and health benefits. Further, as a nation, New Zealand could acquire significant savings in life cycle energy for a modest increase of around 7% in the life cycle cost.

Since LCC involves the future, and the data used are subjected to forecasting, estimation and assumptions, the impact of changes on the results of LCC is likely to vary markedly. Therefore, a sensitivity analysis was carried out to identify the extent to which the results of the above analysis depend on the assumptions used. The following parameters were identified as being subject to risk of change and uncertainty and the range of variation was identified to be within the ranges given below.

- Rate of inflation – 1% to 3% (based on target of Reserve Bank of New Zealand)
- Discount rate – 3% to 9% (based on AS/NZS 4536:1999)
- Price of energy – 20% decrease to 90% increase and 1% annual increase (assumed)
- Period of analysis – 50 years and 75 years (assumed)
- Replacement cycle – high replacement and low replacement (see Table 6.4).

The NPV for the best estimate of events was calculated with no inflation, a discount rate of 5%, 15.95 cents/unit and 72.26 cents/day for energy, 100 years as the period of analysis and best estimate for the replacement cycle as per Table 6.4.

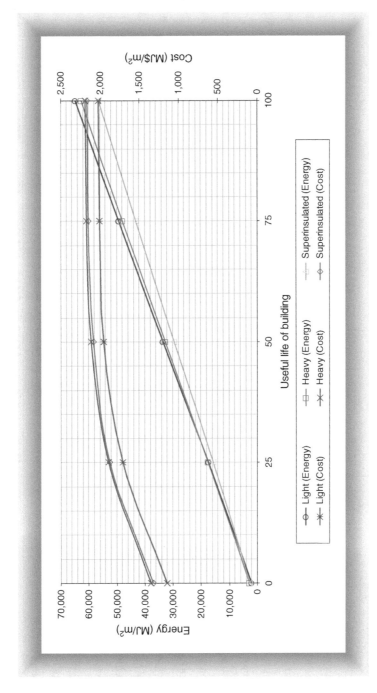

Fig. 7.13 Comparison of life cycle cost and life cycle energy for common constructions (furniture and appliances included).

Table 7.12 Comparison of life cycle energy and life cycle costs for various constructions (furniture and appliances included)

Category	Life cycle energy (MJ/m²)*					Life cycle cost (NZ$/m²)*				
	Year 0	Year 25	Year 50	Year 75	Year 100	Year 0	Year 25	Year 50	Year 75	Year 100
Light construction type										
Construction	1908	2239	3165	3964	4531	973	1100	1207	1224	1228
Furniture & appliances	181	464	773	1095	1261	175	292	333	346	348
Space heating energy#	0	3110	6220	9330	12440	0	57	74	79	80
Other operating energy uses	0	11666	23333	34999	46666	0	259	335	358	365
Total	**2088**	**17479**	**33491**	**49388**	**64897**	**1148**	**1708**	**1949**	**2006**	**2021**
Heavy construction type										
Construction	2793	2925	3464	4023	4356	1177	1292	1374	1395	1397
Furniture & appliances	181	464	773	1095	1261	175	292	333	346	348
Space heating energy#	0	2958	5915	8873	11830	0	54	70	75	76
Other operating energy uses	0	11666	23333	34999	46666	0	259	336	358	365
Total	**2974**	**17669**	**32796**	**47957**	**62736**	**1352**	**1897**	**2113**	**2173**	**2186**
Superinsulated construction type										
Construction	2448	2780	3706	4581	5148	1161	1314	1394	1420	1424
Furniture & appliances	181	464	773	1095	1261	175	292	333	346	348
Space heating energy#	0	1633	3267	4900	6533	0	30	39	41	42
Other operating energy uses	0	11666	23333	34999	46666	0	259	335	358	365
Total	**2629**	**15776**	**29544**	**43272**	**56537**	**1336**	**1894**	**2101**	**2165**	**2178**

*Due to rounding off, values may not add up
line charges excluded from space heating and added to other energy uses

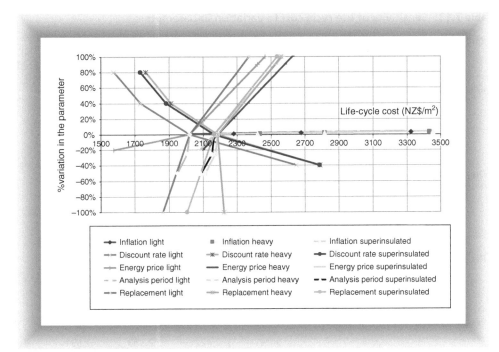

Fig. 7.14   Spider diagram for sensitivity analysis of life cycle cost parameters.

The results of the sensitivity analysis are given in Figure 7.14 and Table 7.13 below.

Figure 7.14 shows that the NPV of life cycle cost is more sensitive to variations in the inflation rate than to any other parameter. This is closely followed by the discount rate. The changes in the parameters, however, do not change the ranking of constructions. It should be borne in mind that, in this analysis, the parameters are varied only one at a time although, in reality, several parameters could vary simultaneously and the final result could be somewhat different.

Use of energy is considered by many researchers to be an indication of the environmental impact due to buildings. However, looking at energy alone does not provide a complete picture of the environmental impacts arising, owing to the numerous and diverse activities related to the building industry. Camilleri (2000a: p. 4) identified greenhouse gases as having the most important impact in terms of NZ houses. The following section considers the greenhouse gases associated with NZ houses and the three construction types.

Table 7.13 Sensitivity of life cycle cost to variation in parameters

Parameter	Estimate	Per cent variation	Life cycle cost (NZ$/m²)		
			Light	Heavy	Superinsulated
Inflation	1%	1%	2,281	2,434	2,421
	2%	2%	2,680	2,817	2,816
	3%	3%	3,325	3,435	3,455
	0%	0%	2,021	2,186	2,165
Discount rate	3%	−40%	2,652	2,789	2,787
	5%	0%	2,021	2,186	2,165
	7%	+40%	1,733	1,911	1,883
	9%	+80%	1,577	1,763	1,732
Period of analysis	50	−50%	1,949	2,113	2,088
	75	−25%	2,006	2,173	2,152
	100	0%	2,021	2,186	2,165
Price of energy		100%	2,466	2,627	2,572
		90%	2,421	2,583	2,531
		40%	2,199	2,362	2,328
		20%	2,110	2,274	2,246
		0%	2,021	2,186	2,165
		−20%	1,577	2,098	2,084
Replacement	High	100%	2,371	2,562	2,538
	Best	0%	2,021	2,186	2,165
	Low	−100%	1,865	2,222	2,006

Life cycle greenhouse gas emissions analysis of the NZ house – model results

Life cycle greenhouse gas (GHG) emissions of various constructions due to use of different building materials were compared with the aim of finding the construction with the least emissions in life cycle terms. A comparison of life cycle embodied GHG emissions for the three construction types is given in Table 7.14.

Floor, walls and roof constitute a major fraction of the mass of a building. As such, they could be expected to contribute extensively to life cycle GHG emissions. In the light and superinsulated construction types considered however,

Table 7.14   Life cycle GHG emission factors for the BIAC standard house

Building element	GHG emission factors (kg CO_2 equiv./m^2)					
	Light construction		Heavy construction		Superinsulated construction	
Foundation	2	1%	6	2%	2	1%
Floor	− 32	− 13%	61	23%	− 28	− 11%
Walls	7	3%	13	5%	− 7	− 3%
Roof	42	17%	− 13	− 5%	51	20%
Joinery	20	8%	20	8%	29	11%
Electrical work	14	6%	14	5%	14	5%
Plumbing	111	44%	111	42%	111	43%
Finishes	85	34%	51	19%	85	33%
Total	**251**	**100%**	**262**	**100%**	**257**	**100%**

floor construction reduced GHG emissions by 13% and 11% respectively. Wall construction of the superinsulated construction type also recorded a 3% reduction in GHG emissions. This is due to carbon locked in timber framing and other timber-based products used in the construction of these elements. Roof construction in the heavy construction type also reduced GHG emissions, by 5%. Roof construction of light and superinsulated constructions however, contributes 17% and 20% respectively to the life cycle GHGs. The figures differ so much because of the emissions due to the use of steel sheets as roof covering on the light and superinsulated houses. The heavy construction house has a roof of concrete tiles, which have much lower emissions, even allowing for the carbon emissions associated with cement manufacture. The high value for the floor of the heavy construction is the result of the cement content of the concrete slab.

Due to shorter replacement cycles during the useful life of the building, finishes are replaced many times and therefore contribute 34%, 19% and 33% of the life cycle building GHG emissions for common, heavy and superinsulated construction types, respectively. Therefore, when considered in life cycle terms, not only the mass of the construction element but the replacement cycle is of utmost importance.

A comparison of life cycle GHG emissions due to materials embodied in the different constructions is given in Figure 7.15.

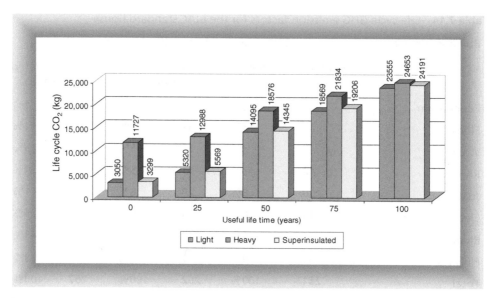

Fig. 7.15   Comparison of life cycle building GHG emissions.

At the initial stage, GHG emissions of the heavy construction type are respectively 3.8 and 3.5 times those of light and superinsulated construction types. All three construction types are, however, similar in terms of life cycle GHG owing to relatively low maintenance requirements for heavy construction. The above comparison of building emissions does not include emissions due to transport and site activities. Emissions due to these activities are estimated to be 6% of total construction emissions. Table 7.15 shows a comparison of life cycle GHG emissions for common construction types over the useful life including the emissions due to space heating energy use.

In terms of life cycle embodied GHG emissions, the value for heavy construction is about 4% higher than that for light construction, while for superinsulated construction the levels are about 2% higher. The CO_2 emissions due to space heating energy use are 86%, 83% and 74% of the life cycle emissions for light, heavy and superinsulated construction types respectively. GHG emissions due to embodied materials and space heating energy use are graphically represented in Figure 7.16.

Life cycle CO_2 emissions follow a pattern similar to that of life cycle energy for the three construction types. While total CO_2 emission for the superinsulated construction is about 8% higher at the initial stage compared to that of the light construction, it is 44% less at the end of the useful life. CO_2 emissions for

Table 7.15 Life cycle GHG emissions for common construction types (excluding furniture and appliances)

Activity	CO_2 equivalent GHG emissions (kg)				
	Year 0	Year 25	Year 50	Year 75	Year 100
Light construction type					
Initial construction	3,050	5,320	14,095	18,569	23,555
Transport & site activities (6% of construction emissions)	183	319	846	1,114	1,413
Total for construction	3,233	5,639	14,940	19,683	24,968
Space heating (2123 kWh per annum)	0	37,136	74,272	111,408	148,544
Total	**3,233**	**42,775**	**89,212**	**131,091**	**173,512**
Heavy construction type					
Initial construction	11,727	12,988	18,576	21,834	24,653
Transport & site activities (6% of construction emissions)	704	779	1,115	1,310	1,479
Total for construction	12,430	13,767	19,690	23,145	26,132
Space heating (2019 kWh per annum)	0	32,304	64,608	96,912	129,216
Total	**12,430**	**46,071**	**84,298**	**120,057**	**155,348**
Superinsulated construction type					
Initial construction	3,299	5,569	14,345	19,206	24,191
Transport & site activities (6% of construction emissions)	198	334	861	1,152	1,451
Total for construction	3,497	5,904	15,206	20,358	25,643
Space heating (1115 kWh per annum)	0	17,840	35,680	53,520	71,360
Total	**3,497**	**23,744**	**50,886**	**73,878**	**97,003**

heavy construction are 284% higher at the initial stage and 10% lower at the end of the useful life. Since CO_2 emissions in NZ houses are mainly due to energy use this pattern might be expected. Figure 7.17 gives a comparison of life cycle CO_2 emissions for the three construction types including space heating energy use.

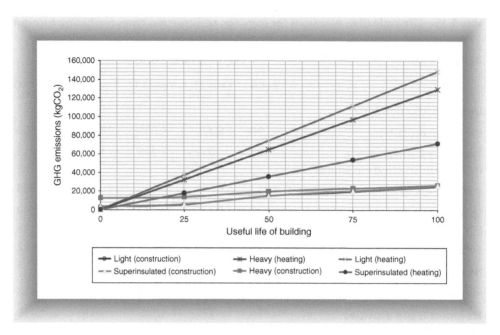

Fig. 7.16 Comparison of GHG emissions due to embodied materials and space heating.

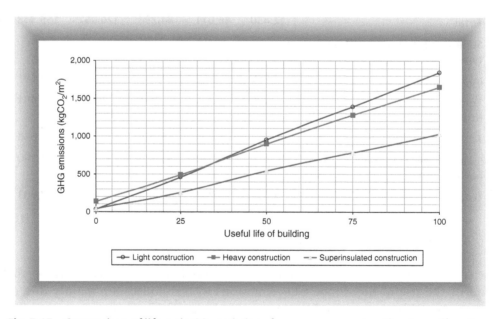

Fig. 7.17 Comparison of life cycle CO_2 emissions for common construction types (furniture and appliances excluded).

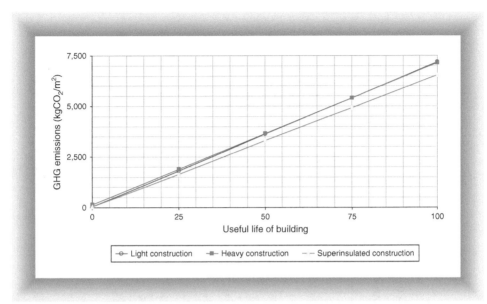

Fig. 7.18 Comparison of life cycle CO_2 emissions for common construction types (furniture and appliances included).

The above comparison does not include emissions due to domiciliary energy uses with the exception of usage of space heating energy. Figure 7.18 shows a comparison of total CO_2 emissions due to all activities within the house including the emissions related to furniture and appliances over the useful life. When total emissions are taken into consideration, at the initial stage heavy and superinsulated constructions emit 217% and 6% more compared with light construction while being 5% and 9% lower respectively at the 100th year.

Environmental impacts due to building construction other than greenhouse gas emissions are not location specific. The following section considers these impacts using the rating schemes derived (see also Chapter 6, Environmental impacts other than GHGs due to construction).

Life cycle environmental impact analysis of an NZ house – model results

Environmental impacts other than greenhouse gas emissions of various constructions due to use of building materials were calculated to find the construction with the least environmental impacts in life cycle terms. A comparison of life cycle environmental impacts for the three construction types is given in Table 7.16.

Table 7.16 Comparison of construction related environmental impacts of the three construction types

Building element	Light construction			Heavy construction			Superinsulated construction		
	Rating	EE*	Impact	Rating	EE*	Impact	Rating	EE*	Impact
Foundation	2	2%	4	3	3%	9	2	2%	4
Floor	1	12%	12	3	28%	84	2	16%	16
Walls	3	25%	75	5	31%	155	4	25%	100
Roof	2	22%	44	1	13%	13	3	24%	72
Joinery	1	13%	13	1	9%	9	1	13%	13
Electrical work	1	5%	5	1	4%	4	1	4%	4
Plumbing	1	9%	9	1	6%	6	1	7%	7
Finishes – floor	4	9%	36	2	5%	10	4	7%	28
– walls/ceiling	2	3%	6	2	1%	2	2	2%	4
Total	–	100%	204	–	100%	292	–	100%	248

* EE = embodied energy

In the comparison given by Table 7.16, the contribution made by various components to the total impact, varies with the construction type. While contribution by the floor construction is only 6% of the total for both light and superinsulated constructions it is 29% for the heavy construction. Walls, roof and finishes contribute 79% of the total for light construction; walls alone contribute 53% of the total for the heavy construction. In contrast, walls, roof and finishes contribute 82% of the total for the superinsulated construction. While the total impact due to superinsulated construction is 22% higher in comparison with the light construction, it is 15% less in comparison with the heavy construction. Therefore, the ranking of construction types based on environmental impact is similar to the ranking based on both life cycle energy and cost. A comparison of the environmental impacts of the three constructions is given in Figure 7.19.

The comparison shown in Figure 7.19, however, does not include space heating energy use and the impacts due to it. Although the comparison made is useful in determining the generic constructions suitable for a certain situation based on the environmental impacts, for a complete picture the impacts due to space heating energy use have to be included in the analysis.

The BIAC house used for the analysis, if constructed according to the energy efficiency code (NZS 4218:1996) requirements, would use 2453 kWh per annum

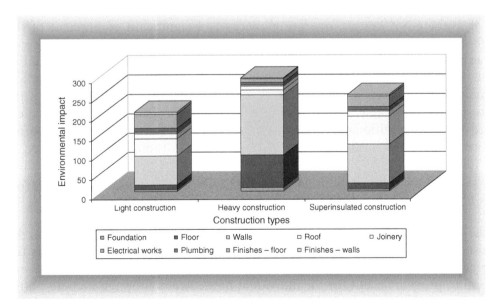

Fig. 7.19   Environmental impact due to material use for various construction types.

as space heating energy. Based on the environmental impact ratings for space heating (see Table 6.10), the space heating energy requirements of light, heavy and superinsulated construction types are rated 6, 5 and 3 respectively. Life cycle impacts of the three constructions with space heating energy use based on the rating system established are as given in Table 7.17.

While the impact due to construction contributes 40% of the total for the superinsulated construction type, it is only 12% and 15% for light and heavy construction types, respectively. Total impact due to superinsulated construction is 44% less compared with the light construction, while it is 34% less compared with the heavy construction. In contrast, the impact due to the heavy construction was only 14% less than the impact due to the light construction. Comparison of life cycle impacts of the three construction types is shown in Figure 7.20.

When the impacts of the space heating requirements are included, those due to construction seem relatively insignificant. However, this gives a more complete picture. While construction impact analysis might aid in selecting the generic construction types that may be suitable, life cycle impact analysis indicates the performance of the whole building. In selecting a suitable construction, total performance has to be considered.

Conclusions

For average NZ houses, operating energy is a significant component of the life cycle energy. Reduction of life cycle energy is not reliant on the use of thermal mass, which is less common in New Zealand due to the requirements of the Earthquake Code, sloping terrain and traditional construction practices. However, if the mass is used it needs to be combined with insulation and passive solar design principles for enhanced performance. Provision of additional insulation does significantly improve the performance of the common light timber framed house. Although houses located in colder climates and those with larger floor areas use more energy over their lives, introduction of additional insulation and double glazing could be used to reduce their overall impact.

The decision to invest or otherwise in additional insulation would depend on the cost. The initial cost increases with the additional insulation and remains the same throughout the useful life. Therefore, the increase in initial cost does not seem to provide benefit for the individual houseowner although it could buffer the owner against any sudden increases in energy prices while providing improved comfort and additional health benefits. As a nation New Zealand

Table 7.17 Comparison of life cycle environmental impacts of various constructions

Building element	Light construction			Heavy construction			Superinsulated construction		
	Rating	% LCE*	Impact	Rating	% LCE*	Impact	Rating	% LCE*	Impact
Foundation	2	0.2	0.4	3	0.5	1.5	2	0.3	0.6
Floor	1	1.3	1.3	3	4.6	13.8	2	3.1	6.2
Walls	3	3.5	10.5	5	5.1	25.5	4	6.2	24.8
Roof	2	4.3	8.6	1	2.1	2.1	3	7.6	22.8
Joinery	1	2.7	2.7	1	2.8	2.8	1	5.1	5.1
Electrical work	1	1.3	1.3	1	1.3	1.3	1	1.9	1.9
Plumbing	1	3.2	3.2	1	3.4	3.4	1	4.6	4.6
Finishes – floor	4	6.7	26.8	2	3.2	6.4	4	8.7	34.8
– walls/ceiling	2	3.8	7.6	2	4.0	8.0	2	6.5	13.0
Space heating	6	73.0	438.0	5	73.0	365.0	3	56.0	168.0
Total	–	**100.0**	**500.0**	–	**100.0**	**430.0**	–	**100.0**	**282.0**

* % LCE = percent of life cycle energy

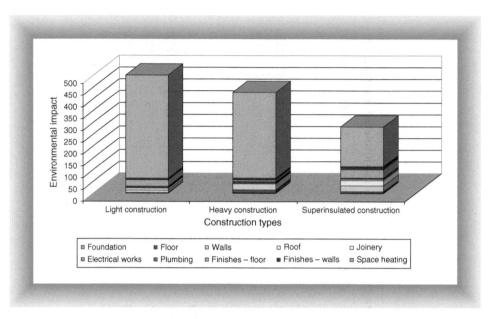

Fig. 7.20 Comparison of life cycle environmental impacts of various construction types (furniture and appliances excluded).

could acquire significant savings by using additional insulation for a marginal increase in life cycle cost.

The associated greenhouse gas emissions and the resultant climate change follow a pattern similar to that of life cycle energy use; the use of additional insulation significantly reduces such emissions. Furniture and appliances make a significant contribution to both life cycle energy and greenhouse gas emissions over the useful life of a building due to their relatively short life. Improvements in their function could be expected to enhance the performance of both new-build and existing houses quite significantly.

For a quick comparative analysis, looking at operating energy values is a useful shorthand means to predict overall environmental impact. It is also a handy way to make a comparative assessment of greenhouse gas emissions.

Notes

[1]Centre for Housing Research

References

Alcorn, A. (1996) *Embodied Energy Coefficients of Building Materials.* Centre for Building Performance Research, Victoria University of Wellington.

Alcorn, J. A. and Haslam, P. J. (1996) The Embodied Energy of a Standard House – Then and Now. In: *Proceedings of the Embodied Energy: The Current State of Play Seminar* (G. Treloar, R. Fay and S. Tucker, eds), pp. 133–140. Deakin University, Australia.

Baines, J. T. and Peet, N. J. (1995) *1991 Input–Output Energy Analysis Coefficients.* Taylor Baines and Associates (for Centre for Building Research Wellington).

Baird, G. and Chan, S. A. (1983) *Energy Cost of Houses and Light Construction Buildings* (Report No. 76), New Zealand Energy Research and Development Committee, University of Auckland.

Bonny, S. and Reynolds, M. (1988) *New Zealand Houses Today.* Weldon.

Breuer, D. (1988) *Energy and Comfort Performance Monitoring of Passive Solar Energy Efficient New Zealand Residences* (Report No. 172), New Zealand Energy Research and Development Committee, University of Auckland.

Camilleri, M. J. (2000a) *Implications of Climate Change for the Construction Sector: Houses* (BRANZ Study Report No. 94). Building Research Association of New Zealand.

Centre for Housing Research (2004a) *Changes in the Structure of the New Zealand Housing Market: Executive Summary.* (Prepared by DTZ New Zealand.)

Centre for Housing Research (2004b) *Housing Costs and Affordability in New Zealand: Executive Summary.* (Prepared by DTZ New Zealand.)

Fay, M. R. (1999) *Comparative Life Cycle Energy Studies of Typical Australian Suburban Dwellings.* PhD Thesis, Deakin University, Australia.

Honey, B. G. and Buchanan, A. H. (1992) *Environmental Impacts of the New Zealand Building Industry* (Research Report 92/2). Department of Civil Engineering, University of Canterbury, New Zealand.

Isaacs, N. (2004) Supply Requires Demand – Where Does All of New Zealands' Energy Go? *Royal Society of New Zealand Conference*, Christchurch, 18 November 2004 (reprint *BRANZ Conference Paper No. 110, 2004*).

Isaacs, N., Camilleri, M., French, L. et al. (2006) *Energy Use in New Zealand Households* (HEEP Year 10 Report). Building Research Association of New Zealand.

Isaacs, N. P. and Trethowen, H. A. (1985) *A Survey of House Insulation* (Research Report R46). Building Research Association of New Zealand.

Jaques, R. (1996) Energy Efficiency Building Standards Project: Review of Embodied Energy. In: *Proceedings of the Embodied Energy: The Current State of Play Seminar* (G. Treloar, R. Fay and S. Tucker, eds), pp. 7–14. Deakin University, Australia.

Mithraratne, N. and B. Vale (2004) Optimum Specification for New Zealand Houses, Paper No. 35 in CD-ROM *Infrastructure and Buildings, Proceedings: International Conference on Sustainability Engineering and Science*, Auckland, 6th–9th July 2004, 14pp.

NZS3604:1999 *Timber Framed Buildings.* Standards New Zealand, 1999.

NZS4218:1996 *Energy Efficiency: Housing and Small Building Envelope.* Standards New Zealand.

Page, I. and Stoecklein, A. (1997) Development of Acceptable Solutions for the Revised Energy Clause of the New Zealand Building Code (Residential Buildings). *IPENZ Annual Conference*, Dunedin, New Zealand, February 1996 (reprint *BRANZ Conference Paper No. 32, 1997*).

Standards New Zealand (n.d.) *Life Cycle Costing: An Application Guide (Australian/New Zealand Standard).* AS/NZS 4536:1999, Standards Australia & Standards New Zealand.

Wilson, D. (2006) *New Zealand Building Economist.* Plans & Specifications Ltd. 35(3).

Wright, J. and Baines, J. (1986) *Supply Curves of Conserved Energy: The Potential for Conservation in New Zealand's Houses.* Centre for Resource Management, University of Canterbury and Lincoln College, New Zealand.

8 Life cycle assessment results and the building user

The previous discussion of life cycle assessment is aimed at the designers of buildings, and also at clients since they pay for buildings. However, a group very influential in relation to life cycle environmental impact, and often ignored, are the users of buildings. The rule of thumb is that how a building works is a third the result of what the designers have achieved, a third how well the building has been constructed and a third down to the users. Over the life of a building the users will dictate how much energy is used to run the building and often the form that energy takes, as they will be the ones paying for it, and they may also dictate the rate at which the building is refurbished and, also, the resources that go into maintaining the building. This chapter seeks to discuss issues that affect life cycle environmental impacts, such as operating performance, finishes, appliances and furniture from the viewpoint of the building user.

Operating energy

It may seem perverse to start with operating energy, since how well a building performs would seem to be tied up with the design, for example whether maximum use has been made of solar gain or whether the building is a highly insulated construction to reduce the rate of heat loss (cold climate) or heat gain (hot climate) and how well the building can be ventilated to remove unwanted heat during hot periods. However, because operating energy is normally the largest part of life cycle impact, understanding how a building has been designed to work is an important part of user behaviour and understanding how to use periodic refurbishment and maintenance to improve operating performance is often up to the user.

Building characteristics

Whatever the climate, a building, such as a house, can be thought of as a mixture of mass materials (e.g. brick wall, mud brick vault, concrete slab) and insulation materials (e.g. fibreglass batts in the wall, thatched roof, expanded polystyrene sandwich cladding panel). How a building performs depends upon the mixture of these materials, but it can be simply thought of in the following way. For the cave dweller, who lived in a mass building with no insulation, the internal temperature would settle at the annual average temperature. For New Zealand that was 13.1°C in 2005, the fourth highest such annual average on record[1] and in the UK it is 8.5–11°C[2]. This explains the old adage that a house with thick stone walls always felt warm in winter and cool in summer, since that was what it was relative to the outside temperature, even if the actual indoor air temperature around 10°C or even 13°C did not represent comfort. Thus, having a lot of mass in a building means the internal temperature will tend to be stable. The presence of insulation in combination with mass will tend to raise the stable internal temperature above the average annual temperature. The experience with Hockerton houses in the UK, which have no space heating system apart from the gains from solar energy, the occupants and the equipment inside, suggests that a 7–8°C temperature rise above the annual average can be achieved with a very high mass construction with 300 mm of insulation to walls, roof and floor, and with the best available off-the-shelf windows. The latter consisted of plantation-grown softwood frames with triple-glazed units, along with krypton gas filling and low-emissivity coatings on two of the glass layers (Vale and Vale 2000: pp. 187–194).

Conversely, the temperature inside a house that has minimal mass and insulation will follow the outside temperature unless energy is put into the house in the form of sunlight, or from the people and equipment housed in it. Temperatures over a day are lowest in the night and highest around midday. In a lightweight house, insulation will lift the internal temperature above the outside temperature. However, the temperature in the house will still go up and down, following the track of the outside temperature but a number of degrees above it. The level of insulation will determine how much the temperature inside is lifted above that outside. For an unheated house in New Zealand with 150 mm of insulation in walls and floor, and 200 mm in the roof, the temperature was lifted about 7°C above that outside when the outside temperature was at its lowest. This meant the minimum indoor temperature recorded in a bedroom was 14°C (Vale and Vale 2001). The windows in this instance were double-glazed with one low-emissivity coating in aluminium frames with no thermal break.

Although it is true to say that in New Zealand the majority of houses are of the lightweight model, in many countries houses are a mixture of mass and light-weight materials, often having masonry walls (mass), concrete slab ground floor (mass), timber joisted upper floor (lightweight) and a timber frame roof (light-weight), and hence their characteristic performance, if they are unheated, will also be somewhere between the two extremes. It is important to have a basic understanding of how buildings might behave. This is because it may fall to the user to attempt to correct any shortcomings in the original design at points of major refurbishment in the building lifetime. During any refurbishment it is unlikely that mass will be added to the building but it is often possible to add insulation. However, before discussing this subject further, it is also necessary to consider the behaviour of small buildings in hot climates and the behaviour of large buildings.

In a hot climate, whether hot wet or hot dry, the aim is usually to keep the building cooler than outside, although in some desert climates where the nights are cold it is also desirable at times to try to raise the inside temperature above that outside. From this it can be seen that in the hot dry desert climate with a large swing in temperature between night and day, the very high mass building is a good solution as it will maintain the annual average temperature. For example, the mean annual temperature in Egypt is $20-25\,^{\circ}\mathrm{C}$[3], which would provide a good comfort temperature in a building. The classic high mass building for this climate was made of mud brick, had few openings to keep out the sun and formed part of a cluster of buildings to keep as much exterior surface as possible shaded from exposure to direct sunlight. In a hot, humid climate the temperature swing day and night and summer to winter is often less, so there is less need of the tempering effect of mass, and the traditional building was often lightweight, and open as much as possible to any cooling breezes to help keep the occupants comfortable. In all hot climates the roof is an important element in keeping out the sun. Often, ventilation paths would be open under the roof in order to keep air flowing over its underside, with the aim of channelling away any heat coming through. Roofs would also be insulated against heat gain. In all warm climates a light-coloured roof is also an advantage to reduce the solar gain into the building.

A small building is dominated by the performance of the surface, walls, roof and floor, as the volume of space enclosed is relatively small. However, a large building has less surface area for the volume enclosed, so its thermal performance tends to be dominated by what happens in it – the gains from people and activities – rather than by the skin, although large areas of glass cladding exposed to the sun will have an effect on internal performance. Large buildings, with one exception, also differ because they tend to be used during the working day and so there is no necessity to maintain comfortable conditions during the night, the time of lowest external temperatures in climates that need heating. The exception is the apartment block, which, especially in Asia, is becoming the norm. The improved performance of such buildings lies more with the designers and constructors as the improvements to the life cycle impact that can be made by the users are limited. Because this chapter is about the effect of the building user, the remainder of the discussion will be centred on the home and the small-scale building.

Building insulation during refurbishment

Life cycle environmental impact includes the energy and resources taken to operate buildings and also the resources that go into periodic maintenance and refurbishment. It is during refurbishment that the major opportunities to improve on the performance of the original design occur. Refurbishment must include repair to make sure that the materials that go into the building continue to function without decay. An example of this type of refurbishment is periodic painting or staining to protect woodwork that would otherwise suffer damage from exposure to the weather. However, more major refurbishment, such as remodelling the building, can give rise to the chance to reduce operating energy through the incorporation of insulation materials. It is also possible to undertake deliberate thermal upgrading of buildings in situations where the energy performance is very poor, to reduce running costs. The simplest example of the latter approach is installation of ceiling insulation to reduce heat loss through the roof. This will not necessarily affect the stability of the temperature in the house, as this type of insulation is of a lightweight element, but it should widen the gap between internal and external temperature, thereby reducing the energy input needed to keep the same degree of comfort. The same approach applies to lightweight walls, such as timber-framed varieties, insulation being able to be placed between the outer and inner coverings. If the inner linings only are removed and replaced, however, the level of insulation is limited to the thickness of the wall framing, normally under 100 mm. If the outer waterproofing cladding is removed, it might be possible to extend the wall structure to include additional layers of insulation, provided the eaves of the building are wide enough such that the top of the thicker wall is still covered. Timber-framed floors that are accessible underneath can be insulated in the same way, or the internal floor covering can be lifted to give access and replaced after the insulation has been installed between the joists. If walls, floor and roof of a lightweight house are insulated, the windows will also need thermal upgrading alongside the rest of the fabric, otherwise they will become very weak spots in terms of heat loss. However, the positive aspect of thermal insulation in a lightweight building that needs heating is that insulation of any part thereof will help to reduce heat loss and therefore will reduce the energy input needed to maintain the same level of internal comfort. However, the same is not necessarily true of a mass building. Thus, it is fair to say that in any building the lightweight elements should be the first to be insulated, and generally this is the easiest task to carry out.

For a mass element, such as a brick or concrete wall, how the building is heated and where the insulation is placed are both critical. If the insulation is on the inner surface of the mass, this will make the building behave as if it were a lightweight building. Adding the insulation exterior to the mass means that some way has to be found to waterproof the insulation and protect it from damage, which may require more secondary structure installed to support some form of waterproof cladding. All this makes the external insulation of existing masonry walls a more expensive and technically difficult undertaking than either internal insulation or the insulating of a lightweight structure. It may also mean extending the eaves of the building to ensure the top of the insulated wall is protected by the eaves. However, external insulation does not destroy the inside of the building and in circumstances where the inside is of historic significance this may be the only possible approach.

The ultimate aim with any scheme of insulation must be to create a building that maintains reasonable, or acceptable, comfort without the need for additional energy input. However, most refurbishments will fall short of this goal, which means the insulation is present to reduce energy usage rather than eliminate it. If heating is more or less continuous, this will be true whether the building is lightweight or has significant mass. Where heating is intermittent however, mass can be a problem as some energy will be needed to heat up the mass each time the heating is switched on. The higher the insulation levels, the less will be the effect, and, for a house that receives good solar gain in the daytime, which will be stored in the mass, with only intermittent morning and evening heating the effect may also not be great. The problem of using extra energy to heat the mass arises only in cases where the lightweight roof may have been insulated but the brick (mass) walls have not been, and heating is intermittent.

From all this a number of guidelines emerge as summarised below:

(1) To make a comfortable building that needs no heating requires adequate mass with adequate insulation on its external face.
(2) Insulation of lightweight elements of a building is relatively simple and for a heated building will reduce energy use.
(3) Mass elements must also be adequately externally insulated if heating is used.

Although this section has focused on adding insulation during refurbishment, it is also possible to add mass. The usual way this is done is by replacing a timber joist floor with a concrete slab. If this mass is also exposed to the sun, it can make a useful contribution to maintaining comfort within the building provided the mass is exposed (such as being tiled) rather than covered with carpet. However, at the time of pouring such a slab the one opportunity of insulating adequately underneath it arises, as it is unlikely it will be dug up in the future in order to add extra insulation. To perform adequately, the floor slab must be insulated.

As insulation is added to walls, roof and even floor, the existing windows become the weak point for heat loss from a small building. Although they can never be insulated sufficiently to make them as effective as wall or roof, thermal upgrading of windows is also important as part of the thermal refurbishment. Looking at windows in terms of life cycle energy, the additional energy taken to make windows with multiple panes, some of which may have low-emissivity coatings, and with heavy gases such as argon or krypton in the gaps between the panes to reduce heat transfer by convection between the panes, will be paid back within a very few years by the energy savings of having more effective windows.

An issue linked to higher-performance windows is the heat that can be lost from the frames unless these are made of a material like timber that is not a particularly good conductor of heat. However, against the energy saved has to be offset the possible additional embodied energy in the form of paint or stain that would be needed to maintain the timber windows over the life of the building. In fact, as it is the operating energy that is the biggest segment of all life cycle energy in houses, every step that can be taken to reduce operating energy use should be taken. Modern metal windows can be obtained with thermal breaks in the frames to eliminate the problem of the frame conducting heat from the warmer interior to the colder exterior. Moreover, ensuring the frames are insulated avoids the problem of moisture condensing on them, especially where the higher specification of the windows means the glass is better insulated than the frame. Condensation on frames can soon lead to mould growth.

Traditionally windows have often been supplied with a series of temporary layers in the form of curtains, blinds and shutters, some of which are used for keeping the sun out while letting the air flow through, while others are used as insulation to keep heat in. A Victorian window, apart from the glass, might have had wooden shutters, and lace curtains, blinds and heavy drapes that would touch the floor to make sure no convection current circulated behind them. These simple ideas can be applied to existing buildings where it is not appropriate, for instance buildings of historical importance, or where it is not economical, to replace single-glazed windows with new energy-saving windows. However, such layers over openings would change the internal appearance and it would move houses away from the Modernist aesthetic of unadorned glass.

Ventilation

The discussion of windows above has concentrated on their thermal performance but windows normally also have the function of opening to provide a means of ventilation. Where there is a need to save energy, at least during some parts of the year, the conventional wisdom is to make the building as air-tight as possible in terms of its construction details, so as to avoid unwanted infiltration of cold air, and then to deliberately ventilate as required. Ventilation is necessary to freshen the internal air, to expel unwanted pollution such as smells from cooking, and perhaps most importantly to get rid of moisture. In occupied houses there is normally more moisture in the internal air than in the external, as living and breathing within a house are moisture-generating activities. Grandmother's wisdom that it was good to sleep with the window open and certainly good to open all the bedroom windows in the morning to air the room, was simply to deal with the problem of the moisture generated when the occupants were asleep at night. Opening the windows would make sure that the moisture was lost to the exterior before it could condense within the room and do damage, either to the building fabric or to health through mould growth. The problem now is that many people are not at home to ventilate the house during the warmest parts of the day, and because the whole family may be out, for example at work and at school during the day, the windows cannot be left open because of the need for security. Consequently, the house may never be ventilated, even though it has opening windows. This has led to a number of strategies, such as installing equipment to deliberately ventilate spaces such as kitchens and bathrooms when moisture-generating activities occur. The simplest method is an extractor fan that is turned on as required. More sophisticated types have a heat exchanger built in so that the incoming fresh air can be warmed by the heat in the outgoing stale air. Both air streams pass through the heat exchanger but are never mixed. In this way, most of the energy in the outgoing air can be transferred to the incoming fresh air. Centralised systems with larger heat-recovery units can be installed, and can provide efficiencies of 80%, thus ensuring adequate fresh air without an energy penalty during the heating season. Such systems do need energy to run the fans required, but the amount of energy required is normally very small. Space has to be found for the ducts. Typical systems will extract from 'wet' rooms, such as kitchens and bathrooms, and supply to living rooms and bedrooms. Such systems can also operate securely when a house is unoccupied.

A simpler solution for daytime secure ventilation is lockable window stays to allow windows to be safely left open. The simplest way to save energy for ventilation, however, is behaviour. People who live 'dry lifestyles', and do not dry washing indoors, or boil vegetables in open pans, will have less moisture indoors and, therefore, less need to ventilate to avoid condensation. The benefits of an unheated conservatory as a secure and safe place to dry washing can be noted, as a conservatory can often warm up quickly during the day when the sun comes out. Fresh air for daytime ventilation can also be taken from an unheated conservatory and may benefit from some pre-warming in this way.

Lighting

In the days when life was sustainable, given that the way of life continued unchanged for many centuries, people tended to get up with the dawn and go to bed at dusk. In this way the sun provided the light required to do the necessary work. In fact those belonging to medieval guilds were forbidden to work at night by candlelight as there would be insufficient light to ensure that the work done was of a high enough standard. Moreover, there have been periods in architectural design when environments seem to have been deliberately designed to encourage ideal conditions for seeing well indoors. Thus, the Georgian facade, with its many equally spaced high and narrow windows, provided a good view of the sky, and hence good natural light to the interior. The windows often had sloping reveals, and were set back in the walls with a gradation of tone between the bright exterior, the white-painted external reveals, and the cream-coloured shutters that provided no sharp contrast in brightness between the highly lit glass and the more dimly lit interior. This all allowed for good seeing, rather than high levels of light. The modern method is to increase the light level to aid seeing, and hence the energy usage is increased. Modern windows, too, are larger and the contrast between the bright glass and the dim interior can often lead to glare. It may be of significance that many modernist interiors were painted white simply because their highly glazed facades were bright, and a white interior was also bright, thus reducing the contrast between the interior and the window. Designing windows which are good for seeing may also link back to the discussion of layers over the glass in terms of shutters, blinds etc. Allowing for adjustable layers allows the user to create the ideal lighting environment for the task in hand, something that is very difficult to do if the window is just an expanse of glass in a thin wall.

This suggests that designers need to rediscover skills in designing environments which are good for seeing, rather than simply trying to design environments with sufficient natural light. Although it may no longer be possible to go to bed with the sun, and hence avoid the need for artificial light, there is a need to reduce energy use through only using sufficient light for the task in hand and through using technologies which can provide more light and waste less energy as heat, such as fluorescents and compact fluorescent lights. In terms of life cycle assessment and conventional economics, although these lamps may have a higher initial cost, because of their long life and the fact less energy is used over their life, they make conventional economic sense. LED technology is coming, possibly allowing for whole surfaces of a room to be illuminated and this technology will reduce the energy needed for lighting even more, but may be of limited use during refurbishments. LED lamps that will fit into conventional sockets and give white light are still expensive. The simplest way to save energy for lighting is not to switch lights on, and to turn them off when they are no longer needed.

Hot water

Generalities about energy are always inaccurate, but they are also useful as a way of seeing the bigger picture. In a 'typical' New Zealand house, for example, very roughly a third of the total energy consumption of 11 410 kWh is used for heating (34%), a third for hot water (29%) and a third (37%) for cooking, lights and appliances (Isaacs et al. 2006: pp. ii and 17). In a UK house, because of the colder climate, the split of the total consumption of 22 795 kWh[4] is 62% for heating, 23% for hot water and 16% for cooking, lights and appliances.

As discussed above, the heating energy needs of a house are more or less fixed once it is built, and may be hard to change. Mass and insulation last the life of the house. However, hot water systems wear out and may be replaced more than once over the lifetime of a house. For hot water systems that rely on non-renewable energy inputs, it is probably preferable to use an on-demand heating system, in order to avoid having a large tank of water which is constantly losing heat. The 'standing loss' from a hot water cylinder may be more than 25% of its total energy consumption. However, if solar energy is used to heat water, storage becomes essential so that there is enough hot water available when the sun is not shining. The thickness of insulation around the hot water cylinder and on the pipe supplying hot water to the house will determine how long the water keeps hot.

It is possible to reduce hot water demand through simple technical measures, such as low-flow shower heads and in-line flow reducers for taps, but another way to reduce the energy needed for hot water is through behavioural change. Some people have 20-minute showers, whereas others take 2 minutes to shower, but in both cases they manage to keep clean. Some people rinse their dishes under running hot water while others wash up in a 5 litre bowl of hot water.

Appliances

Although hot water systems may be replaced once or twice over the lifetime of a house, appliances are often replaced very frequently. This is partly to do with the cycle of fashion – many people move into a house and immediately replace the existing kitchen – and partly because appliances break down and it is often cheaper to buy new than to have the old ones repaired. People also have many more appliances in their homes than once was the case, and electrically powered versions have replaced non-electric versions. For example, alarm clocks used to be clockwork driven, but are now either mains or battery powered. In the kitchen, electric devices have replaced the traditional balloon whisk and the later geared beaters. Very occasionally these replacements save energy; for example, making a loaf of bread in a breadmaker uses only 0.4 kWh, whereas baking it in a conventional oven will require 0.8 kWh. However, use of the majority of kitchen appliances means increased energy consumption.

Sometimes appliances break down without anyone realising, and this can have severe energy penalties. For example, the 10 year Household Energy End-Use Project (HEEP) research in New Zealand found that around 7% of refrigerators and freezers had faulty thermostats, so were never shutting off (Isaacs et al., 2006: p. iii). It is not easy to recognise if your refrigerator is faulty, and the result could be high energy consumption. The fridge/freezer, in any case, is traditionally one of the most energy-hungry of appliances: some, even when operating correctly, may use more than 1000 kWh per year, or, maybe, 10% of a household's total energy consumption. A 290 litre fridge freezer which used 760 kWh in 1980 uses only 254 kWh in 2005 under the European A++ energy rating (CECED 2006).

New appliances are not necessarily more efficient than older ones. Plasma televisions use more energy than cathode ray tube televisions, but of course the plasma screens are larger. The energy used per square inch of the screen is not necessarily greater, but the desire to have the home-theatre experience is increasing home energy consumption, as can be seen in Table 8.1. The energy used by a family's television viewing could be as little as 130 kWh per year, or as much as 1450 kWh, meaning that it might vary roughly between 1.5 and 15% of total household energy consumption in New Zealand.

All appliances need to be chosen to be as low energy as possible, but users also need to consider behaviour in using appliances. For example, it is clear from Table 8.1 that some televisions use a lot of energy in stand-by mode, and turning the television off could save 200 kWh per year in the worst case scenario. Similarly, buying an energy-saving washing machine is a good idea, but using cold-water detergent would also save energy, because the water would not have to be heated.

Embodied and maintenance energy

Finishes

With regard to the consideration of embodied energy, the lowest-energy finishes are the ones that would be thought of as traditional, such as limewash, whitewash and beeswax polish. The history of modern finishes is one of those that need as

Table 8.1 Energy consumption of a range of televisions

Type	Model	Size (in.)	Power (W)				Annual energy (kWh*)
			TV	DVD	Standby	per sq. in.	
LCD	Samsung SyncMaster 151MP	15	43.9	43.7	11.8	0.41	200
LCD	Sharp LC-20B8U-S	20	18.4	45.2	5.7	0.16	130
CRT	RCA 27F634T	27	86.7	85.6	2.5	0.25	270
CRT	Sharp 27DV-S100	27	124.9	92.7	3.5	0.32	340
LCD	Envision A27W221	27	104.9	103	5.1	0.37	330
LCD	JVC LT-32X776	32	114.2	129	11.1	0.24	420
LCD	ViewSonic N3250w	32	152	148	3.9	0.31	460
CRT	Sony KD-34XBR960	34	189.1	209	5.2	0.4	610
LCD	Sony KDL-V40XBR1	40	214.4	212.9	24	0.31	760
Plasma	Hitachi 42HDT52	42	360	205	37	0.39	1040
Plasma	Maxent MX-42X3	42	357.6 **	256	17.9	0.3	1000
Plasma	Maxent MX-50X3	50	414.2 **	414	16.8	0.39	1310
Plasma (power save off)	Panasonic TH-50PHD8UK	50	236.1 **	332.1	16.1	0.27	920
Plasma (power save on)	Panasonic TH-50PHD8UK	50	229.2	280.3	16.1	0.24	840
Projector	Panasonic PT-52LCX65	52	172	172	6.9	0.15	540
Plasma	Hitachi 55HDT52	55	434.1	507.1	13.2	0.36	1450
Projector	Sony KDS-R60XBR1	60	223	220	10.8	0.14	710
Projector	JVC HD-61FH96	61	202	199	3.4	0.11	610
Projector	Mitsubishi WD-62628	62	235	235	39.8	0.14	920
Projector	HP MD6580n	65	276	274	10.3	0.15	860
Projector	Samsung HL-R6768W	67	231	229	27.4	0.12	830

* Based on 4 hours of DVD, 4 hours of TV, and 16 hours of standby
** Did not contain tuner; tested via RF input connected to external tuner displaying EPG
Based on: http://reviews.cnet.com/4520-6475_7-6400401-3.html?tag=arw (accessed 4 Jan 2007)

little short-term maintenance as possible, but these come with an energy penalty. Thus, a polyurethane varnish applied to timber will save the need to polish it regularly and will need nothing more than an occasional wash, but polyurethane is both energy-intensive in its manufacture and potentially harmful to the user (most polyurethane finishes have the label 'Poison' on the container).

During refurbishment, choosing finishes that are as natural as possible and not oil-based, will tend to reduce environmental impact, even if these finishes have to be replaced more frequently. The advantage of using old-fashioned finishes is that they do not pollute the indoor atmosphere as many oil-based finishes do. The smell that is associated with new paints, or even buying a new car, is indicative of off-gassing of chemicals, often poisonous, into the indoor air. However, the need to polish the floor more frequently because a natural wax is used is another change in user behaviour.

Furniture

There was a time when furniture was so valuable that records exist of it being handed down from one generation to the next (Quiney 1986: pp. 45–9). However, life cycle analysis has shown that the short life of modern furniture means it can have a considerable effect on the overall life cycle impact of the building if it is included as part of the analysis. The change in attitude to furniture would appear to have come with the Industrial Revolution, with its creation of a wealthy middle class who wanted to have houses furnished in imitation of the wealthy aristocracy, using affordable products made by machine. This led to a number of outcries at the time against the general standard of design, and in part gave rise to the Arts and Crafts Movement with its emphasis on the hand-made, and therefore, expensive, as typified by the furniture of the Barnsley Brothers and Gimson. To be fair, the Arts and Crafts interiors were sparse as regards to the attitude to furnishing, suggesting that furniture was something expensive and to be prized, as had happened in the past. At the same time, designers associated with this movement also introduced the idea that furniture should be designed to go with the building interior, and also with the overall exterior and landscaped setting. These ideas permeated down from the wealthy – who had always followed design fashion in the main rooms of their great houses even if the older but still useable furniture continued to be used in the less frequented rooms of the house – to the middle classes. Interiors at the new garden suburb at Hampstead were shown fitted out with suitable furniture, as were many of the Parker and Unwin middle-class houses that were built in Derbyshire at the start of the twentieth century (Parker and Unwin 1901: Plates 44, 56, 58, 60). This moved furniture from being something that most people used second hand to fashion item status.

The problem with furniture being a fashion item was two-fold. First, it introduced the idea that furnishings should be changed relatively frequently. In turn, this implied that furniture should not be constructed to last for centuries as had happened in the past, because, if the chair or cupboard fell apart after a few years of use no problem would arise since it would no longer be in fashion. Only when resources were scarce, such as in Britain during World War II, was there recognition of the fact that design and manufacture should be improved and that furniture could be designed to use minimal resources and still last. This approach was part of the Utility design movement in the UK and the design of Utility furniture was very much led by Gordon Russell, himself a designer of furniture with an Arts and Crafts pedigree (ILEA 1974: pp. 3, 25–26). The manufacturers at first resisted the idea but later welcomed the careful designs that produced furniture with minimal resources and which lasted well. The designs were very much in the vernacular furniture tradition but streamlined by contact with Modernism and the exigencies of minimising resource use.

In the present, only old furniture which has earned the right to be called antique is really valued, and here it is rarity that is being purchased rather than a particular style or type of furniture; nevertheless, the idea of valuing the old is important in terms of preserving resources and will become increasingly so in a sustainable future.

Guidelines on dealing with the issues

This section aims to suggest guidelines and priorities for the users of buildings to help reduce life cycle environmental impact.

- As operating energy is the largest component of life cycle energy, its reduction should be the priority, rather than worrying about embodied energy.
- Insulation is generally the key to reducing operating energy and refurbishment should see insulation as a priority. It is generally easiest to insulate the light-weight elements of a building first and, for a heated building, this will reduce energy use. Mass elements must also be adequately externally insulated even if heating is used.
- To make a comfortable building that needs no heating requires adequate mass with adequate insulation on its external face.
- Once the fabric of the building has been insulated, the windows should also be upgraded. Adding layers to windows in the form of blinds, shutters and curtains is a simple approach to improving window performance.
- Avoid having a very 'wet' lifestyle, to avoid damage from moisture in the home.

- If it is not possible to ventilate the house in the daytime by opening the windows, some other form of deliberate ventilation system should be used to remove moisture from the interior of the house.
- Switching off a light when it is not needed is the simplest way to save energy.
- Compact fluorescent lamps, although more expensive to buy, do make sense in life cycle terms both for life cycle energy and life cycle cost.
- Taking shorter showers is the quickest way to save the energy used to heat water.
- If a hot water system needs replacing, a solar water heating system with an adequate storage tank might be an option to reduce the life cycle environmental impact of a hot water supply.
- Install Energy Star and European A-class rated appliances where these are available.
- Turn off appliances at the wall whenever possible.
- Use natural finishes rather than those based on petroleum products.
- Use of second-hand or antique furniture will reduce the overall life cycle environmental impact.

Conclusion

None of the issues presented in the guidelines above should be a surprise as they will be found in many discussions on how to make houses and other small buildings use less energy and have less impact on the natural environment. What life cycle analysis allows is the chance to set priorities, as it is possible to see precisely what contributes to the making of the life cycle impact and the relative size of the constituent parts. If there is one thing life cycle analysis confirms, it is the importance of insulation in reducing life cycle energy use and life cycle impact. Insulation is a boring subject, as the money spent on it is generally not visible in the way money spent on a state-of-the-art kitchen is. Nevertheless, at every stage of a building's life its environmental performance will be improved by the addition of insulation. Perhaps it is the case that building designers as well as users have to learn to love increased levels of 'invisible' insulation within their homes.

Notes

[1] http://www.niwascience.co.nz/ncc/cs/aclimsum_05
[2] http://www.metoffice.com/climate/uk/location/england/#temperature
[3] http://www.fao.org/ag/AGL/swlwpnr/reports/y_nf/egypt/e_clim8.htm

[4]http://www.statistics.gov.uk/STATBASE/ssdataset.asp?vlnk = 7287&More = Y; conversion from 1960 kg oil equivalent using conversion value of 11,630 kWh per tonne of oil equivalent from: http://www.carbontrust.co.uk/resource/ energy_units/default.htm

References

CECED (2006) *Energy Efficiency; A Shortcut to Kyoto Targets; The Vision of European Appliance Manufacturers.* European Committee of Manufacturers of Domestic Equipment (CECED).

Inner London Education Authority (1974) *Utility Furniture and Fashion 1941–1951*, ILEA.

Isaacs, N., Camilleri, M., French, L. et al. (2006) *Energy Use in New Zealand Households* (HEEP Year 10 report). Building Research Association of New Zealand.

Parker, B. and Unwin, R. (1901) *The Art of Building a Home*. Longmans Green.

Quiney, A. (1986) *House and Home*. BBC.

Vale, B. and Vale, R. (2000) Thermal Mass in Zero-Heating Houses. In: *Renewable Energy Transforming Business: From Fossils to Photons, Solar 2000 Proceedings* (D. Mills, J. Bell, L. Stoynov and P. Yarlagadda, eds). *38th Annual Conference of the Australian and New Zealand Solar Energy Society*, Brisbane, 29 November – 1 December 2000.

Vale, B. and Vale, R. (2001) Thermal Performance of Superinsulated Lightweight Residential Construction in the Auckland Climate. Paper NOV09 in CD-ROM *Performance in Product and Practice: Proceedings: CIB World Building Conference*, Wellington, 2–6 April, 9pp.

Appendix A: Frequency of occurrence of common specifications

Element	Item	Description	%
Foundation		Tanalised timber piles	21%
		House piles (timber/concrete)	31%
		Reinforced concrete footings	30%
Floor		Timber-framed suspended floor with particleboard flooring	65%
		Concrete slab floor	35%
	Floor insulation	Double-sided aluminium foil draped over floor frame	58%
		Polystyrene perimeter insulation	8%
External walls		Kiln dried light timber frame	60%
		Double tongue & grooved laminated timber	29%
		Precast insulated concrete panels	7%
		Steel frame	4%
	Wall insulation	Glass fibre batts	58%
		Polystyrene	7%
		Double sided aluminium foil	7%
	Wall cladding	Fibre cement weather board	47%
		Brick veneer	13%
		Timber cladding	9%
		Fibre cement backing board with a finishing coat	2%
Internal walls		Timber frame with plasterboard	65%
		Prefabricated wood-based panel	8%

Roof	Frame	Timber truss	53%
		Timber rafters & beams	36%
		Steel truss	11%
	Covering	Corrugated steel	87%
		Concrete tiles	12%
	Insulation	Glass fibre	86%
Ceiling		Plasterboard	56%
		Wood fibre tiles	30%
Joinery	Windows	Aluminium	99%
	Front door	Glazed aluminium	52%
		Timber panel	43%
		Steel	5%
Plumbing	Gutters & down pipes	Metal	47%
		PVC	36%

Appendix B: Embodied energy coefficients of New Zealand building materials

Material		Alcorn 1996		Alcorn and Wood 1998		Alcorn 2003	
		MJ/kg	MJ/m³	MJ/kg	MJ/m³	MJ/kg	MJ/m³
Aggregate	general	0.1	150	0.1	150	0.04	65
	virgin rock	0.04	63	0.04	63	0.06	83.3
	river	0.02	36	0.02	36	0.03	46.7
Aluminium	virgin	191	515,700	191	515,700	192	517,185
	extruded	201	542,700	201	542,700	202	544,685
	extruded, anodised	227	612,900	227	612,900	226	611,224
	extruded, factory painted	218	588,600	218	588,600	218	587,940
	foil	204	550,800	204	550,800		
	sheet	199	537,300	199	537,300		
Aluminium	recycled	8.1	21,870	8.1	21,870	9	24,397
	extruded	17.3	46,710	17.3	46,710	14.6	39,318
	extruded, anodised	42.9	115,830	42.9	115,830	23.8	64,340
	extruded, factory painted	34.3	92,610	34.3	92,610	15.2	40,928
	foil	20.1	54,270	20.1	54,270		
	sheet	14.8	39,960	14.8	39,960		
Asphalt (paving)		3.4	7,140	3.4	7,140	0.2	335

Material		Alcorn 1996		Alcorn and Wood 1998		Alcorn 2003	
		MJ/kg	MJ/m³	MJ/kg	MJ/m³	MJ/kg	MJ/m³
Bitumen	fuel	44.1	45,420	44.1	45,420	44.3	45,632
	feedstock					2.4	2,475
Brass		62	519,560	62	519,560		
Carpet		72.4		72.4			
	felt underlay	18.6		18.6			
	nylon	148		148			
	polyester	53.7	7,710	53.7	7,710		
	polyethyltere-phthalate (PET)	107		107			
	polypropylene	95.4	57,600	95.4	57,600		
	wool	106		106			
Calcium carbonate				1.3			
Cellulose pulp				14.3		19.6	1,057
Cement	average	7.8	15,210	9	17,550	6.2	12,005
	dry process			7.7	15,020	5.8	11,393
	wet process			10.4	20,280	6.5	12,594
	cement grout			1.9	4,560		
	cement mortar	2	3,200	2.1	3,360		
	fibre cement board	9.5	13,550	10.9	15,550	9.4	13,286
	soil–cement	0.42	819	0.7	1,420		
Ceramic	brick, new technology	2.5	5,170	2.5	5,170	2.7	5.310
	brick, old technology			7.7	1,580	6.7	13,188
	brick, glazed	7.2	14,760	7.2	14,760		
	brick, refractory			5.7	12,825		
	pipe	6.3		6.8	13,880		
	tile	2.5	5,250	2.5	5,250		

Material		Alcorn 1996		Alcorn and Wood 1998		Alcorn 2003	
		MJ/kg	MJ/m³	MJ/kg	MJ/m³	MJ/kg	MJ/m³
Concrete	block	0.94		0.94		0.9	12.5/ unit
	block-fill			1.4	3,150	1.2	2,546
	block-fill, pump mix			1.5	3,430	1.2	2,732
	brick	0.97		0.97			
	GRC	7.6	14,820	7.6	14,820		
	grout			1.7	2,380	1.5	3,496
	paver	1.2		1.2			
	pre-cast	2	2,780	2	2,780	1.9	4,546
Ready mix conc.	17.5 MPa	1	2,350	1	2,350	0.9	2,019
	17.5 MPa pump mix			1.2	2,830		
	30 MPa	1.3	3,180	1.3	3,180	1.2	2,762
	40 MPa	1.6	3,890	1.6	3,890	1.4	3,282
	roofing tile	0.81		0.81			
Copper	virgin	70.6	631,164	70.6	631,160	2.4	21,217
	virgin, sheet					97.6	872,924
	virgin, rod, wire					92.5	827,316
	recycled, tube					2.4	21,217
Earth, raw	adobe block, straw stabilised	0.47	750	0.22	360		
	adobe, bitumen stabilised	0.29	490	0.29	490		
	adobe, cement stabilised	0.42	710	0.67	1,130		
	clay			0.07	45		
	clay for cement			0.1	65		
	rammed soil cement	0.8	1,580	0.73	1,450		
	pressed block	0.42	810	0.42	840		

Material		Alcorn 1996		Alcorn and Wood 1998		Alcorn 2003	
		MJ/kg	MJ/m³	MJ/kg	MJ/m³	MJ/kg	MJ/m³
Fabric	cotton	143		143			
	polyester	53.7	7,710	53.7	7,710		
Glass	general	15.9	37,550				
	float	15.9	40,060	15.9	40,060	15.9	40,039
	toughened	26.2	66,020	26.2	66,020	26.4	66,605
	laminated	16.3	41,080	16.3	41,080	16.3	41,112
	tinted	14.9	375,450	14.9	375,450	15.9	40,039
Insulation	cellulose	3.3	112	3.3	110	4.3	146
	fibreglass	30.3	970	30.3	970	32.1	1,026
	polyester	53.7	430	53.7	430		
	polystyrene, expanded	117	2,340	117	2,340	58.4	1,401
	wool (recycled)	14.6	139	20.9	200		
Lead		35.1	398,030	35.1	398,030		
Linoleum		116	150,930	116	150,930		
Paint		90.4	117,500 (118/l)	90.4	117,500 (118/l)		
	solvent based	98.1	127,500 (128/l)	98.1	127,500 (128/l)		
	water based	88.5	115,000 (115/l)	88.5	115,000 (115/l)		
Paper		36.4	33,670	36.4	33,670		
	building	25.5		25.5			
	kraft	12.6		13.9			
	recycled	23.4		23.4			
	wall	36.4		36.4			
Plaster, gypsum		4.5	6,460	3.8	5,480	3.6	8,388
Plaster board		6.1	5,890	6.1	5,890	7.4	7,080

Material		Alcorn 1996		Alcorn and Wood 1998		Alcorn 2003	
		MJ/kg	MJ/m³	MJ/kg	MJ/m³	MJ/kg	MJ/m³
Plastics	ABS	111	125,430	111	125,430		
	high-density polyethelene (HDPE)	103	97,340	103	97,340	51	48,166
	low-density polyethelene (LDPE)	103	91,800	103	91,800	51	45,872
	polyester	53.7	7,710	53.7	7,710		
	polypropylene	64	57,600	64	57,600		
	polystyrene, expanded	117	2,340	117	2,340	58.4	1,401
	polyurethane	74	44,400	74	44,400		
	PVC	70	93,620	70	93,620	60.9	80,944
Rubber	natural latex	67.5	62,100	67.5	62,100		
	synthetic	110	118,800	110	118,800		
Sand		0.1	232	0.1	230	0.1	232
Sealants and adhesives	phenol formaldehyde	87		87			
	urea formaldehyde	78.2		78.2			
Steel,	recycled	10.1	37,210	10.1	37,210		
	reinforcing, sections	8.9	69,790	8.9	69,790	8.6	67,144
	wire rod	12.5	97,890	12.5	97,890	12.3	96,544
Steel,	virgin, general	32	251,200	32	251,200	31.3	245,757
	galvanised	34.8	273,180	34.8	273,180		
	imported, structural	35	274,570	35.9	281,820		
	stainless, average			50.4	395,640	74.8	613,535
Stone, dimension	local	0.79	1,890	0.79	1,890		
	imported	6.8	17,610	6.8	17,610		
Straw, baled		0.24	30.5	0.24	30		

Material		Alcorn 1996		Alcorn and Wood 1998		Alcorn 2003	
		MJ/kg	MJ/m³	MJ/kg	MJ/m³	MJ/kg	MJ/m³
Timber, softwood	air dried, roughsawn	0.3	165	0.3	170	2.8	1,179
	air dried, roughsawn, CCA-treated					3	1,252
	kiln dried, roughsawn	1.6	880				
	air dried, dressed	1.16	638	1.16	640	3	1,273
	kiln dried, average, dressed	2.5	1,380	5.09	2,204		
	kiln dried, gas fired, dressed			8.2	3,550	9.5	3,998
	kiln dried, gas fired, dressed, CCA-treated					9.7	4,060
	kiln dried, waste fired, dressed			3.1	1,340	4.1	1,732
	mouldings, etc.	3.1	1,710	3.1	1,710		
	hardboard	24.2	21,300	24.2	13,310		
	MDF	11.9	8,330	11.9	8,330	11.9	8,213
	glulam	4.6	2,530	4.6	2,530	13.6	5,727
	particle board	8	4,400	8	4,400		
	plywood	10.4	5,720	10.4	5,200		
	shingles	9	4,930	9	4,930		
Timber, hardwood	air dried, rough sawn	0.5	388	0.5	390		
	kiln dried, roughsawn	2	1,550	2	1,550		
Vinyl flooring		79.1	105,990	79.1	105,990		
Water	reticulated			0.003	3.3		
Zinc		51	364,140	51	364,140		
	galvanising, per kg steel	2.8		2.8			

Appendix C: CO_2 equivalent greenhouse gas emission factors for New Zealand building materials

Material		This analysis		Alcorn 2003	
		kg/kg	kg/m^3	kg/kg	kg/m^3
Aggregate	general	0.006	8.580	0.0023	3.5
	virgin rock	0.002	3.604	0.0031	4.6
	river	0.001	2.059	0.0016	2.4
Aluminium	virgin	13.306	35,943	8.0000	21,600
	extruded	13.878	37,488	8.3540	22,555
	extruded, anodised	15.365	41,503	9.3590	25,270
	extruded, factory painted	14.851	40,113	9.2050	24,855
	foil	14.050	37,951		
	sheet	13.764	37,179		
Aluminium	recycled	1.059	2,862	0.6220	1,679
	extruded	1.585	4,283	0.7210	1,946
	extruded, anodised	3.049	8,237	0.8870	2,393
	extruded, factory painted	2.557	6,909	0.7310	1,975
	foil	1.745	4,716		
	sheet	1.442	3,897		
Asphalt (paving)		0.194	408	0.0146	22.80
Bitumen	fuel	3	2,598	3.020	3,111
	feedstock			0.171	176

Material		This analysis		Alcorn 2003	
		kg/kg	kg/m³	kg/kg	kg/m³
Brass		4	29,719		
Carpet		4.141			
	felt underlay	1.064			
	nylon	8.466			
	polyester	3	441		
	polyethylterephthalate (PET)	6.120			
	polypropylene	5	3,295		
	wool	6.063			
Calcium carbonate		0.764			
Cellulose pulp				0.612	33
Cement	average	1.00	1,700	0.994	1,939
	dry	0.93	1,555	0.967	1,885
	wet	1.08	1,856	1.021	1,990
	cement grout		743		
	cement mortar		639		
	fibre cement board	0.955	1,327	0.629	894
	soil–cement	0.120			
Ceramic	brick, new technology	0.143	296	0.138	271
	brick, old technology, avg.	0.440	90	0.518	1,021
	brick, old technology, coal			0.684	1,348
	brick, old technology, gas			0.353	695
	brick, glazed	0.412	844		
	pipe	0.389	794		
	refractory brick	0.326	734		
	tile	0.143	300		

Material		This analysis		Alcorn 2003	
		kg/kg	kg/m³	kg/kg	kg/m³
Concrete	block			0.106	1.60
	block fill		486	0.156	345
	block fill, pump mix		479	0.163	375
	brick		330		
	GRC		848		
	pre-cast		449	0.214	526
	grout		194	0.209	496
Ready mix	17.5 MPa		178	0.114	268
	17.5 MPa pump mix		296		
	30 MPa			0.159	376
	40 MPa			0.189	452
	roofing tile		125		
Copper	virgin	40.383	36,102	7.738	69,173
	virgin, rod, wire			7.477	66,844
	recycled, tube			0.112	1,002
Earth, raw	adobe, straw stabilised	0.126	206		
	adobe, bitumen stabilised	0.166	280		
	adobe, cement stabilised	0.464			
	clay	0.040			
	clay for cement	0.057			
	rammed soil cement	0.498			
Glass	float	0.909	2,291	1.735	4,372.00
	toughened	1.499	3,776	1.918	4,834.00
	laminated	0.932	2,350	1.743	4,391.00
	tinted	0.852	21,476	1.626	40,975.22
Insulation	cellulose	0.1888	6.29	0.140	4.70
	fibreglass	1.7332	55.48	0.770	24.60
	polyester	3.0716	24.60		

Material		This analysis		Alcorn 2003	
		kg/kg	kg/m³	kg/kg	kg/m³
	polystyrene, expanded	6.6924	134	2.495	59.90
	wool (recycled)	1.195	11.44		
Lead		2.008	22,767		
Linoleum		6.635	8,633		
Paint		5.171	6,721		
	solvent based	5.611	7,293		
	water based	5.062	6,578		
Paper		2.082	1,926		
	building	1.459			
	kraft	0.795			
	recycled	1.338			
	wall	2.082			
Plaster, gypsum		0.907		0.218	501.00
Plaster board		0.349	337	0.421	404.00
Plastics	ABS	6.349	7,175		
	high-density polyethelene (HDPE)	5.892	5,568	3.447	3,257.00
	low-density polyethelene (LDPE)	5.892	5,251	3.540	3,186.00
	polyester	3.072	441		
	polypropylene	3.661	3,295		
	polystyrene, expanded	6.692	134	2.495	59.90
	polystyrene, extruded	3.340	107	2.495	79.80
	polyurethane	4.233	2,540		
	PVC	4.004	5,355	4.349	5,784.00
Rubber	natural latex	3.861	3,552		
	synthetic	6.292	6,795		
Sand		0.006	13	0.007	15.90
Sealants and adhesives	phenol formaldehyde	4.976			
	urea formaldehyde	4.473			

	Material	This analysis		Alcorn 2003	
		kg/kg	kg/m^3	kg/kg	kg/m^3
Steel	recycled	1.068			
	reinforcing, sections	0.999		0.352	2,766.00
	wire rod	1.205		0.526	4,129.00
Steel	virgin, general	3.790		1.242	9,749.00
	galvanised	3.951	31,012		
	imported, structural	4.013			
	stainless, average	4.843		5.457	44,747.00
Stone, dimension	local	0.045			
	imported	0.389			
Straw, baled		0.014			
Timber, softwood	air dried, roughsawn	− 2.115	− 907	− 1.665	− 699
	kiln dried, roughsawn air dried, dressed	− 2.065	− 880	− 1.662	− 698
	air dried, roughsawn, treated CCA	− 1.960	− 845	− 1.657	− 696
	kiln dried, dressed	− 1.841	− 791	− 1.349	− 567
	kiln dried, gas fired, dressed	− 1.663	− 714		
	gas dried, dressed, treated CCA			− 1.342	− 564
	bio dried, dressed	− 1.954	− 840	− 1.644	− 690
	mouldings, etc.	− 1.954	− 819		
	hardboard	− 0.748	− 155		
	MDF	− 1.451	− 440	− 0.568	− 392.00
	glulam	− 1.869	− 772	− 1.141	− 479.00
	particle board	− 1.674	− 665		
	plywood	− 1.537	− 619		
	shingles	− 1.617	− 635		
Timber, hardwood	air dried, rough sawn	− 2.103	− 894		
	kiln dried, roughsawn	− 2.017	− 828		

Material		This analysis		Alcorn 2003	
		kg/kg	kg/m³	kg/kg	kg/m³
Vinyl flooring		4.525	6,063		
Vitreous china	(based on Baird & Chan 1983)	1.373			
Water	reticulated		0.189		
Zinc		2.917	20,829		
	galvanising, per kg steel	0.160			

Index

T - #0404 - 101024 - C245 - 246/189/13 - PB - 9780750680639 - Gloss Lamination